Practical Manual Series - 4

Biotechnology

NIPA® GENX ELECTRONIC RESOURCES & SOLUTIONS P. LTD.
New Delhi-110 034

www.nipaers.com
Browse, Search, Read & Buy...
Print Books
eChapters
eBooks
Publishers
Forthcoming
Subject Catalogues
New Titles
Online Resources on:
✓ Current Affairs
✓ Reasoning, Logic & Aptitude
✓ Competitive Examinations
✓ English Language Lab
✓ Writing & Pronunciation Tools
✓ Personality Development
✓ Online Programmes for Professionals
✓ Interview Preparation
Pay Using
PhonePe
Google Pay
Paytm
PayPal
UPI
CC Avenue

Practical Manual Series - 4

BIOTECHNOLOGY

By

K. M. THARA

Department of Biotechnology
University of Calicut, CU Campus, Thenjippalam
Malappuram District, Kerala - 673 635 India

NIPA® GENX ELECTRONIC RESOURCES & SOLUTIONS P. LTD.
New Delhi-110 034

NIPA® GENX ELECTRONIC
RESOURCES & SOLUTIONS P. LTD.

101,103, Vikas Surya Plaza, CU Block
L.S.C. Market, Pitam Pura, New Delhi-110 034
Ph : +91 11 27341616, 27341717, 27341718
E-mail: newindiapublishingagency@gmail.com
www: www.nipabooks.com

For customer assistance, please contact
Phone: + 91-11-27 34 17 17
Fax: + 91-11- 27 34 16 16
E-Mail: feedbacks@nipabooks.com

ISBN: 978-81-19002-99-3

Composed and Designed by NIPA®.

PREFACE

Biotechnology is a developing branch of science. It makes use of all basic science for industrial, agricultural and medical level applications. In this Laboratory manual almost all areas of basic Biotechnology are included. It has eight different sections with experimental protocols. Protocols are given in a simplified manner and easy to perform. General information about a Biotechnology laboratory is available in this book. Hope this will serve as a first hand laboratory manual for graduate, postgraduate and research students.

Areas covered in this book include Biochemistry, Microbiology, Enzymology, Plant Biotechnology, Animal Biotechnology, Molecular biology and Genetic engineering. Protocols given in this manual are of basic experiments. Even though it is easy to perform, I request all those who do the experiments will make sure to take maximum care for the best results. Results may vary according to the laboratory conditions, quality of chemicals etc. So a careful standardization is required for good results. It is necessary to follow all general laboratory precautions before conducting all experiments.

I would like to thank all in the Departments of Life Sciences and Biotechnology, University of Calicut, who have supported and encouraged me for writing this book. I thankfully acknowledge those great personalities who have pioneered and developed the various protocols in a useful manner. I humbly remember and thank my teachers who have provided the basic knowledge and guidance throughout the course of my studies. I may dedicate this book to Highness Kollur Sri. Mookambika Devi for the successful completion of this book and everlasting inspirations.

C.U. Campus
1-02-2009

K.M. THARA

CONTENTS

Section-3 : Biochemistry

Section-4 : Enzymes

Section-5 : Microbiology

Section-6 : Plant Biotechnology

Section-7 : Animal Biotechnology

Section-8 : Molecular Biology

Chapter 1

Different Units used in Measurements

1. Length and Size

In Biotechnology measurement of size of organisms and materials which are relatively small are to be determined. The units of measurements used for length and size are in microns at the light microscopic level, and in nanometer at the electron microscopic level. For molecular measurements, the norm is the Angstrom. These units can be given as below.

Measure	Symbol	Relative Length	Exponential Notation
Meter	M	1.0	10^{0}
Decimeter	dm	0.1	10^{-1}
Centimeter	cm	0.01	10^{-2} M
Millimeter	mm	0.001	10^{-3} M
Micrometer or micron	µ	0.000001	10^{-6} M
Nanometer	nm	0.000000001	10^{-9} M
Angstrom	Å	0.0000000001	10^{-10} M
Pico	p	0.000000000001	10^{-12} M

From above relationships it can be given that

10Å=1nm

1000nm=1mm

10 mm = 1 cm

1inch=2.54cm=25.4mm=25,400μ=25,400,000nm

1inch=2.54cm=2.54x10^1mm=2.54x10^{4-}=2.54x10^7nm

1 mm = 0.04 inches

2. Volume

Volumes are measured in liters, and the most commonly used measurements are the milli liters and the micro liters. The following gives the relative volumes.

Measure	Symbol	Relative Volume	Exponential Notation
Liter	L	1.0	10^7
Deciliter	dl	0 .1	10^{-1}
Millimeter	ml	0.001	10^{-3}
Micro liter	μ l	0.000001	10^{-6}

1,000μl=1ml.

1ml=1 cm^3

1gallon=3.8liters

1quart=0.95liters

1 ounce = 29.6 ml

3. Weight

The most common measurements of weights are in grams, milligrams and micrograms.

Measure	Symbol	Relative Weight	Exponential Notation
Kilogram	Kg	1000g	10^3 g
Gram	g	1.0g	10^1 g
Milligram	mg	0.001	10^{-3} g
Microgram	μ g	0.000001	10^{-6} g

Chapter 2

Solution Preparation

1. Concentration

Concentrations can be expressed as Molarity, Molality or Normality. It can also be expressed as weight per volume or percentage.

Molarity : Molarity is based on the number of moles of solute in the solvent. Molarity equals the number of moles of solute in 1 liter of solution. A mole is equal to the gram molecular weight (or formula weight) of the solute. Sodium Chloride (NaCl), for example has a formula weight of 58.43 (22.98 for Na and 35.43 for Cl). Thus, if 58.43 grams of NaCl are dissolved in 1 liter of water, it will be resulted in a 1 molar solution of NaCl. This is designated as 1 M NaCl, or as simple M NaCl.

On dealing with solutions of volume less than 1 liter, milliliter is used for the measurement and as the following may be noted:

1 M NaCl = 58.43 grams / liter = 58.43 mg/ml = 58.43 µg/µl.

A 0.002 M NaCl solution contains 0.002 moles of NaCl or 0.1168 grams (0.002 x 58.43) in one liter of solvent. Note that molar is abbreviated as M, but that there is no abbreviation for moles. The 0.002 M solution contains 0.002 moles (or 2 milli moles) or solute in one liter. A 0.002 M solution would contain 0.001 moles of solute in a half liter.

The number of moles = Volume (in liters) x Molar Concentration

The number of milli moles = Volume (in ml) x Molar Concentration

Note that chemical equations are always balanced via moles. Moreover, note that for dilutions of known concentrations, one can use the simple formula:

Molarity x Volume = Molarity x Volume i.e. M1V1= M2V2

To convert a 0.002 M solution of NaCl to 100 ml. of a 0.001 M solution,

0.002MxNeededVolume,X=0.001Mx100ml

X=0.001Mx0.1liters/0.002M=0.050liters=50ml.

Measure 50 ml of the 0.002 M solution, and dilute it to 100 ml with the solvent (usually water, or an appropriate buffer).

Molarity is appropriate when chemical equations are to be balanced. When physical properties of solutions are measured, molality is used as a similar measurement of concentration. For colligative properties of solutions like freezing point depression, boiling point elevation, osmotic pressure, density, viscosity, there is a better co-relation between the property and molality.

Molality (m): is equal to the number of moles of solute in 1000 gm of solvent. This may appear as same as molarity, since one ml of water equals 1.0 gm. Indeed, for dilute solutions in water, there is little or no practical difference between a molar solution and a molal solution. In concentrated solutions, with temperature fluctuations and with changes in solvent, there is appreciable difference.

Normality (N)

Normality (N) is another measure of concentration that relates the amount of solute to the total volume of solution.

It is defined as the number of equivalents of solute per liter of solution

Normality = number of equivalents of solute / 1 L of solution.

There is a very simple relationship between normality (N) and molarity (M)

$N = n \times M$ (where n is an integer).

For an acid solution, let n is the number of H+ provided by a formula unit of acid.

Example : A 3M H_2SO_4 solution is the same as a 6N H_2SO_4 solution.

For a basic solution, n is the number of OH- provided by a formula unit of base.

Example : A 1 M Ca $(OH)_2$ solution is the same as a 2N Ca $(OH)_2$ solution. For example: A 2 m (2 molal) solution of sucrose contains 684.4 gm of sucrose (twice the molecular weight or two moles of sucrose) dissolved in 1000 gm (approximately 1 liter) of water. The weight of this solution is 684.4 gm + 1000 gm i.e. 1684.4 gm.

Percentage

Percentage is based on the number of parts, either grams (for a solid solute) or milliliters (for a liquid solute) by weight or volume.

Example: 2 m sucrose, there were 684.4 gm of sucrose in the final solution which weighed 1684.4 gm (684.4 gm sucrose + 1000 gm water). The percentage of sucrose on the basis of weight is therefore 684.4/1684.4 x 100, or 40.6%.

There are three means of expressing concentration in the form of a percent figure:

1. Percent by weight (w/w); gm solute / 100 gm solvent
2. Percent weight by volume (w/v); gm solute /100 ml solvent
3. Percent by volume (v/v); ml solute / 100 ml solution

For dilute solutions, these differences are not significant, but at higher concentrations, they are significant. Chemists (when they use Percent designations) usually use w/w. Biochemists and physiologists more often use w/v. Both use v/v if the solute is a liquid. It is important to distinguish among these alternatives.

Using ethanol as an example, consider a 20% solution of ethanol in water, mixed according to the three designations of w/w, w/v and v/v.

1. w/ w would contain 20 g of absolute ethanol mixed with 80 gm of water to yield a 20% (w/w) solution.
2. w/ v would contain 20 g of absolute ethanol mixed with water to form a final volume of 100 ml.

3. v/ v would contain 20 ml of absolute ethanol diluted to 100 ml with water.

The three solutions are not the same. First, the density of alcohol is not equal to that of water, and thus conversion of g to ml is not equivalent. A 20% (w/w) solution of ethanol, for example, has a density of 0.97 g/ml and 20 gm of ethanol plus 80 gm of water would have a volume of 103 ml. The % (w/v) for this solution would be 20 gm ethanol / 103 ml, or 19.4% (w/v). Similarly, absolute ethanol has a density of 0.79 gm/ml and thus 20 ml of ethanol would weigh 15.8 gm. A 20% (v/v) solution would contain 15.8 gm of ethanol in 100 ml and be a 15.8% (w/v) solution.

So, for ethanol:

20% (w/w) =19.4% (w/v)

20% (w/v) = 20.0 % (w/v)

20% (v/v) = 15.8 % (w/v)

Note : Unless specifically stated otherwise, solutions lacking the appropriate designation should be assumed to be (w/v).

ppm = parts per million =mg/liter

Grams/milliliter = g/ml = milligrams/μl = mg/μl

1μg /ml = 1mg/l = 1ppm

ppm = μg/g =μg/ml = ng/mg = pg/ug = 10^{-6}

ppm = mg/litres of water

1g / 1000ml = 1000 ppm

PPB = Parts per billion = ng/g = ng/ml = pg/mg = 10^{-9}

Making up 1000 ppm solutions

1. From a salt of the metal

e.g. make a 1000 ppm standard of Na using the salt NaCl.

Formula Weight (FW) of salt = 58.44g.

Atomic weight of Na = 23

1g Na in relation to FW of salt = 58.44 / 23 = 2.542g.

Hence, weigh out 2.542g NaCl and dissolve in 1 liter volume to make a 1000 ppm Na standard.

2. From an acidic radical of the salt

e.g. Make a 1000 ppm phosphate standard using the salt KH_2PO_4

FW of salt = 136.09

FW of radical PO_4 = 95

1g PO_4 in relation to FW of salt = 136.09 / 95 = 1.432g.

Hence, weigh out 1.432g KH_2PO_4 and dissolve in 1 liter volume to make a 1000 ppm PO_4 standard.

Dilutions

Dilution Formula, N 1 V1 = N 2 V2

ppm of the stock x volume of the stock= required ppm x required volume of the solution to be made

e.g. for obtaining 50 ml volume of 25 ppm from 100 ppm.

25 x 50/ 100 = 12.5 ml. i.e. 12.5 ml of 100 ppm in 50 ml volume will give a 25 ppm solution.

Serial dilutions

Dilution of 10^{-1} M to 10^{-5} M solutions from a 1M stock solution.

Pipette 10 ml of the 1M stock into a 100 ml volumetric flask and make up to100ml mark to give a 10^{-1} M solution.

Now, pipette 10 ml of this 10^{-1} M solution into another 100 ml flask and make up to the mark to give a 10^{-2} M solution.

Pipette again, 10 ml of this 10^{-2} M solution into yet another 100 ml flask and make up to mark to give a 10^{-3} M solution.

Pipette a 10 ml of this 10^{-3} M solution into another 100 ml flask and make up to mark to give a 10^{-4} M solution.

And from this 10^{-4} M solution pipette 10 ml into a 100 ml flask and make up to mark to give a final 10^{-5} M solution.

Molarity to ppm

e.g. Calculate the ppm concentration of calcium in 0.01M $CaCO_3$

At. Wt. of Ca = 40 and 1M = 40 x 1000 = 40,000ppm.

Hence, 0.01M solution = 40,000 x 0.01 = 400 ppm.

The FW of an ion species is equal to its concentration in ppm at 10^{-3}M. Fluoride has a FW of 19, hence a 10^{-3}M concentration is equal to 19ppm, 1M is equal to 19,000 ppm and 1ppm is equal to 5.2 x 10^{-5}M.

Conversion of ppm (parts per million) to % (parts per hundred).

Example

1 ppm = 1/1,000,000 = 0.000001 = 0.0001%
10 ppm = 10/1,000,000 = 0.00001 = 0.001%
100 ppm = 100/1,000,000 = 0.0001 = 0.01%
200 ppm = 200/1,000,000 = 0.0002 = 0.02%
5000 ppm = 5000/1,000,000 = 0.005 = 0.5%
10,000 ppm = 10000/1,000,000 = 0.01 = 1.0%

Parts per hundred- % to ppm
Example: 0.01% = 0.0001
0.0001 x 1,000,000 = 100 ppm

ppm (parts per million) to % (parts per hundred)

2 ppm = 12/1,000,000 = 0.000002 = 0.0002%

20 ppm = 20/1,000,000 = 0.00002 = 0.002%

100 ppm = 100/1,000,000 = 0.0001 = 0.01%

500 ppm = 500/1,000,000 = 0.0005 = 0.05%

5000 ppm = 5000/1,000,000 = 0.005 = 0.5%

10,000 ppm = 10000/1,000,000 = 0.01 = 1.0%

Section 2

BIOTECHNOLOGY LABORATORY

Chapter 3

General Precautions and Requirements

1. General Precautions

Plan and prepare for every laboratory exercise before coming to the laboratory.

Keep work bench neat and clean. Wipe with detergents if necessary.

Do not eat, drink, smoke, chew gum, or apply cosmetics in the laboratory. Never use a laboratory microwave, laboratory, fridge or freezer for food. Never eat or drink using laboratory glasswares.

Dispose of biological materials such as microbes by sterilizing and in the special containers for biological hazards. Do not throw other materials into these bags.

Dispose of broken glass and other sharp materials in the separate disposal boxes labeled "sharp materials". Do not throw materials other than "sharps" into the "sharp materials" receptacles.

Never use cracked or broken glasswares.

Be sure you know where to dispose of hazardous chemicals. Do not wash solvents or hazardous materials down the drains. Do not combine chemicals for disposal unless you know they are not reactive with one another.

Wear proper attire. Wear a laboratory coat. Sometimes gloves are also required.

Always wear safety glasses or other eye protection.

Tie back long hair.

Be very careful with Bunsen burners. Do not leave a burner lit unless you are standing next to it because someone else might be injured if they do not realize it is lit.

Wash your hands before leaving the laboratory, even if you wore gloves.

Locate eyewash stations, fire extinguishers, and safety showers.

Never mouth pipette. This technique is very dangerous and will result in illness and death. If you used this method in the past, learn to use pipetting aids.

Be familiar with the chemicals in the laboratory and take maximum care in handling each chemical. Keep chemicals back in position after use. Do not spill out chemicals.

Leave coats, extra books, and personal items in your locker. Avoid tripping over these items and taking pathogenic, toxic, radioactive, and other hazardous materials to home and your families.

Clean work bench before leaving the laboratory.

2. General requirements for Biotechnology laboratory

Generally a Biotechnology laboratory should have the following rooms.

1) *Store room*: For storing chemicals and glassware.
2) *Cleaning and washing room*: For general cleaning purpose.
3) *Preparation and sterilization room: For media preparation and media sterilization.*
4) *General laboratory*: For routine laboratory experiments.
5) Specialized rooms for special works like, for
   i) *Animal biotechnology*: The room should have the equipments for the animal biotechnology work specifically like Laminar flow, incubator, water bath, pH meter, balance, osmometer, centrifuge, microscopes, carbon dioxide incubator, disposable sterile culture flasks, bottles, syringes, filter units, micropipettes. Room should be kept clean and sterile.
   ii) *Plant Biotechnology*: Preparation and sterilization room, Laminar flow, sterile storage and culture rooms.
   iii) *Microbial Biotechnology*: Laminar flow, incubator, water bath, pH meter, microscope
6) *General instrumentation room*: For PCR machine, gel documentation system, electrophoresis units, centrifuge, pHmeter, balance, laminar flow, freezers, ice machine etc.
7) Proper disposal of media, cultures etc. Microbial cultures should be sterilized before disposal.
8) Animal room facility with general experimental animals like mice, rat, rabbit, hamsters. Prior permission must be obtained from animal ethics committee before conducting experiments.

Chapter 4

Equipments

1. Equipments

List of general equipments required is given below

1) Weighing balance Different types like
   i) Top loading electronic balance
   ii) Micro Balance for microgram weighing i.e. weight upto 0.0001 mg (Different models available)
2) *pH meter* : Different models available. A pH meter should be maintained according to manufactures instructions. pH meter should be calibrated with standard buffers regularly for accurate results.
3) *Magnetic stirrers*: For proper mixing of solutions
4) *Autoclaves*: Different models like horizontal, vertical, with different capacities are available.
5) *Laminar air flow*: Different models available. Generally horizontal or vertical model. UV should be kept on

½ an hour before the work. Laminar flow should be kept clean and dust free.

6) *Bacteriological incubators:* For incubating cultures at specific temperatures

7) *CO_2 incubators* : Mainly used for animal cell culture. Incubators are connected to CO_2 cylinders and the rate of CO_2 flow, humidity, temperature etc should be controlled according to the requirements.

8) *Microscope* : Different types of microscopes are available
   i) Ordinary clinical microscope-Magnification up to 100x, general laboratory purpose.
   ii) Phase contrast
   iii) Inverted phase contrast
   iv) Electron microscope

9) Spectrophotometer/calorimeter

10) Centrifuge
   i) Clinical centrifuge
   ii) Ultracentrifuge

11) Electrophoresis unit: Different models like
   i) Horizontal-for DNA Electrophoresis
   ii) Vertical for proteins
   iii) 2D gel electrophoresis for amino acids

12) *Gel documentation system* : For viewing DNA/protein gels, and analysis.

13) *PCR machine* : For DNA amplification work-Different models and types are available
    RT-PCR-Real Time PCR machines

14) *Ultra sonicator* : For homogenization of tissue-The instrument should be kept in sound proof cases as it can cause damage to the ears.

15) *ELISA reader* : For ELISA plate reading

16) *Hybridization oven:* For hybridization in blotting work

17) Freezers (-20° C - 70 °C)

18) *Cold room panel:* Maintains a temperature between 0-10, Use generally for extraction, storing and processing purposes.

19) Liquid nitrogen cans and liquid nitrogen for cryo preservation, freezing and homogenization purposes.

20) *DNA Sequencer-Automatic* : For sequencing the gene

21) *Fraction collector* : For programmed collection of fraction in column chromatography

22) *HPLC:* Different models available

23) *Shakers:* For cultures, mixing purposes

24) *Hot air oven:* For drying and sterilization purpose.

25) *Water bath*: For incubation at specific temperature

26) Laboratory fermenter with pH, temperature and air supply controls.

27) Water purification systems for
    i) Double distilled water
    ii) Sterile water

28) TLC set

29) Ice making machine

30) *Lyophilizer* : For freeze drying of samples

Chapter 5

Principles of Some Common Instruments

1. Centrifuge

When a particle move in a circular path a force is applied towards the centre of the circle along the radius and this force is known as centrifugal force. The use of centrifugal force to separate some biologically important substances was initiated by T. Svedberg and J.W Williams in the year 1920. They coupled mathematics, physics and mechanics for developing the technique of centrifugation.

The sedimentation velocity which is the required velocity in a centrifugal field to sediment the particle is known as its Svedberg constant or s value. Centrifugation involving quantitative work is known as ultracentrifugation.

Relative centrifugal force (RCF) : Centrifugal force is depend upon the radius of the rotation of the rotor ,the speed at which it rotates, size, shape and type of the rotor

.Size and speed can be kept constant at certain point, but the radius varies from top to bottom of the centrifuge tube. So the measurement can be taken from the middle point or as average radius. All forces are mathematically related to gravity and correlates to Relative Centrifugal Force (RCF).

RPM revolutions per minute : RPM vary with a particular instrument and the rotor used. RCF is constant and independent of the instrument used.

RCF= 1.12 X r $(RPM/1000)^2$; r= radius in millimeters, RPM- Revolutions Per Minutes.

If the density and viscosity of the medium and the particles present are known, the time needed for the sedimentation of the particle can be calculated by the formula

Time, T= D-L/D+L * $(N/d_2\ (g\text{-}p)\ S_2)$

T=Time in minutes, D-radial distance in cm(max), L-Radial distance to meniscus, N-viscosity of fluid medium, g=density of fluid medium, p-density of the particle to sediment, d-diameter of the particle in cms, S-rotational velocity in RPM

Rotors : There are two types of rotors which are commonly used.

i) Fixed angle
ii) swinging bucket.

Depending on the nature of the particle to sediment, the type of the rotor has to be selected.

Types of Centrifuges

i) *Clinical*: Ordinary use Speed up to 5000 rpm
ii) *Microfuge*: up to 10,000 speeds, for small volumes
iii) Ultracentrifuges up to 20,000 and above speed.

Types of Centrifugation

i) *Differential centrifugation*: Depending upon the rpm/ speed different fractions can be precipitated and separated according to their mass.

ii) *Density gradient centrifugation*: The particles are separated on the basis of their density. The mixture or homogenate are placed on a mixture of solution of different density or density gradient.

Precautions

i) Different models of centrifuges are available and it should be operated according to the manufacturers instructions.

ii) Every use of ultracentrifuges must be recorded in the log book.

iii) Rotors should be in good condition.

iv) Never use alkali detergent on rotors.

v) Use proper centrifuge tube according to the use.

vi) Fill the tubes properly and close it.

vii) Balance the rotor properly to avoid mainly the bending of rotors. Balance with similar fluids. Not like alcohol with water or sucrose.

viii) Ensure that rotor is properly seated and tightened.

ix) Set time, speed, rotor number etc as per the instrument model.

x) Open the centrifuge only after the rotor stops completely.

2. Spectrophotometer

Principle : When light passes through a medium light will be absorbed by the particles in the medium and the transmitted light will have an intensity lesser than the incident light (Beer's law).

i.e. $I = I0 * 10^{-kcl}$

I = Intensity of the light transmitted

I_0 = Intensity of the incident light

C = Concentration of the solution

l = Distance traveled by the light through the solution

k = Constant

Then the logarithmic relation can be taken as

$-\log I = \log I_0 - kcl$

$-\log l = \log 1/l = k\ c$ = optical density (OD)

OD is directly proportional to the concentration of the colored compound and it is logarithmic scale and % transmittance is an arithmetic scale.

Table: 2.3.2 : The absorption of light of particular wavelength and the color of solutions

Wavelength (nm)	Light absorbed by the object	Color of the object appears to be
380-430	Violet	yellow
430-475	Blue	orange
475-495	Greenish blue	red-orange
495-505	Bluish green	orange-red
505-555	Green	red
555-575	Yellowish green	violet-red
575-600	Yellow	violet
600-650	Orange	blue

Various Parts of a Spectrophotometer

A spectrophotometer consists of a spectrometer as source of monochromatic light of required wavelength and a photometer for measuring the intensity of the light. The sample liquid is placed in a cuvette between the spectrometer and photometer. Light from spectrometer passes through the cuvette with sample liquid and photometer measures the intensity of transmitted light to the display device normally a galvanometer.

Note : Colorimeter is also used for measuring OD. It consists of different filters for fixed wavelength selection while spectrophotometer can be used for a range of wavelengths and it is more accurate and sensitive. Spectrophotometer consists of a prism with grating for light of required wavelength instead of filters in colorimeter.

Precautions

i) Cuvette should be kept clean and dry.
ii) Small packets of silica gel must be kept in the cuvette holder to adsorb the moisture.
iii) Spectrophotometer/Colorimeter should be kept on 20 min in advance to the measurement.
iv) Set the required wavelength according to the manufactures instruction.
v) Keep blank in the cuvette and adjust the reading to zero.
vi) Take sample and note OD.
vii) Blank is the buffer plus reagents.

3. Microscope

Several kinds of microscopy are available. Two major categories are

i) Light microscopy
ii) Electron microscopy.

Light microscopy : Light microscopy makes use of a system of optical lenses, through which light wave passes and magnification occurs. It can be light field, dark field, fluorescence, phase contrast. Fluorescence and confocal microscopes are specialized instruments, used for research, clinical, and industrial applications.

Other than the compound microscope, a simpler instrument for low magnification use can also be found in a laboratory. The stereo microscope or dissecting microscope usually has a binocular eyepiece tube, a long working distance, and a range of magnifications typically from 5x to 40x.

Bright Field Microscopy

In bright field microscopy, light from an incandescent source passes towards a lens from beneath the stage through the condenser, then through the specimen, the objective lens, and to the eyes through a second magnifying lens called the ocular or eyepiece. A magnification up to 1500 x and a 100-200nm resolution will be obtained in bright field microscopy. It is necessary to stain the specimen. The objects are seen in the light path because stains absorb light differentially, or because they are thick enough to absorb a significant amount of light despite being colorless. A good quality microscope has a built-in illuminator, adjustable condenser with aperture diaphragm (contrast) control, mechanical stage, and binocular eyepiece tube. The condenser is used to focus light on the specimen through an opening in the stage. The magnification of the image is simply the objective lens magnification times of the ocular magnification.

Some condensers are fixed in position; others are focusable, so that the quality of light can be adjusted. Usually the best position for a focusable condenser is as close to the stage as possible. The bright field condenser usually contains an aperture diaphragm, a device that controls the diameter of the light beam coming up through the condenser, so that when the diaphragm is stopped down the light comes straight up through the center of the condenser lens and contrast is high. When the diaphragm is wide open the image is brighter and contrast is low.

Mount the Specimen on the Stage

Take a clean dry slide. Mount the material to be observed as a small drop. Stain with the suitable stain .Cover with a cover slip. Place the slide on the stage of the microscope. The stage may be equipped with simple clips or with some type of slide holder. The slide may require manual positioning, or there may be a mechanical stage that allows precise positioning without touching the slide.

Optimize the Lighting

A light source should have a wide dynamic range, to provide high intensity illumination at high magnifications, and lower intensities so that the user can view comfortably at low magnifications. Better microscopes have a built-in illuminator, and the best microscopes have controls over light intensity and shape of the light beam. If your microscope requires an external light source, make sure that the light is aimed toward the middle of the condenser. Adjust illumination so that the field is bright without hurting the eyes.

Adjust the Condenser

For adjusting and aligning the microscope, start by reading the manual.

Focus, Locate, and Center the Specimen

i) Start with the lowest magnification objective lens, to home in on the specimen and/or the part of the specimen you wish to examine. It is rather easy to find and focus on sections of tissues, especially if they are fixed and stained, as with most prepared slides. However it can be very difficult to locate living, minute specimens such as bacteria. A suspension of yeast cells makes a good practice specimen for finding difficult objects.

ii) Start with the specimen out of focus so that the stage and objective must be brought closer together. The first surface has to come into focus as you bring stage and objective together on the top of the cover slip. With smears, a cover slip is frequently not used, so the first thing you see is the smear itself.

If you are having trouble, focus on the edge of the cover slip or an air bubble, or something that you can readily recognize. The top edge of the cover slip comes into focus first, then the bottom, which should be in the same plane as your specimen.

iii) Once you have found the specimen, adjust contrast and intensity of illumination, and move the slide around until you have a good area for viewing.

iv) Adjust eyepiece separation, focus.

v) Select an objective lens for viewing

The lowest power lens is usually 3.5 or 4x, and is used primarily for finding specimens initially. We sometimes call it as scanning lens for this reason. The most frequently used objective lens is the 10x lens, which gives a final magnification of 100x with a 10x ocular lens. Typical high magnification lenses are 40x and 97x or 100x. The latter two magnifications are used exclusively with oil in order to improve resolution.

Move up in magnification by steps. Each time you go to a higher power objective, re-focus and re-center the specimen. Higher magnification lenses must be physically closer to the specimen itself, which poses the risk of jamming the objective into the specimen. Be very cautious when focusing.

Adjust Illumination for the Selected Objective Lens

Bright field microscopy is best suited to viewing stained or naturally pigmented specimens such as stained prepared slides of tissue sections or living photosynthetic organisms.

Magnifications required generally are as below

i) Prepared slides, stained - bacteria (1000x),

ii) Thick tissue sections (100x, 400x),

iii) Thin sections with condensed chromosomes or specially stained organelles (1000x),

iv) Large materials (100x).

v) Smears, stained - blood (400x, 1000x), negative stained bacteria (400x, 1000x).

vi) Living preparations (wet mounts, unstained) - (40x, 100x, 400x)

Care of the Microscope

Good quality microscope is highly expensive, so be careful.

Hold a microscope firmly by the stand only. Never grab it by the eyepiece holder.

Hold the plug (not the cable) when unplugging the illuminator.

Since bulbs are expensive, and have a limited life, turn the illuminator off when you have completed your work.

Always make sure the stage and lenses are cleaned before putting away the microscope.

Never use a paper towel, a Kim wipe or any material other than good quality lens tissue or a cotton swab (must be 100% natural cotton) to clean an optical surface. Clean gently. Appropriate lens cleaner or distilled water should be used to remove dried material from the lens. Organic solvents may separate or damage the lens elements or coatings.

Cover the instrument with a dust jacket when not in use.

Focus smoothly : Don't try to speed through the focusing process or force anything. For example if you encounter increased resistance when focusing then you've probably reached a limit and you are going in the wrong direction.

Dark Field Viewing

Dark field optics is a low cost alternative to phase contrast optics. Usually staining is not required in dark field microscopy. Specimen appears bright on a dark background. A dilute suspension of yeast cells can be used for good dark field optics.

Principle

To view a specimen in dark field, an opaque disc is placed underneath the condenser lens, so that only light that is scattered by objects on the slide can reach the eye. Instead of coming up through the specimen, the light is reflected by particles on the slide. Everything is visible regardless of color, usually bright white against a dark background. Pigmented objects are often seen in "false colors," that is, the reflected light is of a color different than the color of the object. Better resolution can be obtained using dark field as opposed to bright field viewing.

To set up a dissecting microscope for "dark field" viewing, the specimen should be placed over an opening so that light reflects only from surfaces between cover slip and slide, not from a surface beneath the slide. You may need to make a stand to hold the slide. The surface beneath the opening should be a flat black. Turn off any built-in illuminator. Aim a high-intensity light source toward the specimen at an angle, from the top or side through a glass dish or jar.

Dark field illumination is most readily set up at low magnifications (up to 100x), although it can be used with any dry objective lens. Any time you wish to view everything in a liquid sample, debris and all, dark field is best. Even tiny dust particles are obvious. Dark field is especially useful for finding cells in suspension. Dark field makes it easy to obtain the correct focal plane at low magnification for small, low contrast specimens. Use dark field for initial examination of suspensions of cells such as yeast, bacteria, or cell and tissue fractions including cheek epithelial cells, chloroplasts, mitochondria, even blood cells (small diameter of pigmented cells makes it tricky to find them sometimes despite the color). First locate any specimen of very small size and try to focus the required specimen.

Phase Contrast Microscope

Most of the detail of living cells is undetectable in bright field microscopy because there is too little contrast between structures with similar transparency and there is insufficient natural pigmentation. However the various organelles show wide variation in refractive index, that is, the tendency of the materials to bend light, providing an opportunity to distinguish them. This change in refractive index can be converted in to the change in phase of the reflected light and can be viewed differentially under the lense. When light passes through from one material to another material which is of slightly different refractive index / thickness will undergo a change in phase. This can be converted into difference in brightness and can be detected.

Principle

Highly refractive structures bend light to a much greater angle than do structures of low refractive index. The same properties that cause the light to bend also delay the passage of light by a quarter of a wavelength or so. In a light microscope in bright field mode, light from highly refractive structures bends farther away from the center of the lens than light from less refractive structures and arrives about a quarter of a wavelength out of phase.

Light from most objects passes through the center of the lens as well as to the periphery. Now if the light from an object to the edges of the objective lens is retarded a half wavelength and the light to the center is not retarded at all, then the light rays are out of phase by a half wavelength. They cancel each other when the objective lens brings the image into focus. A reduction in brightness of the object is observed. The degree of reduction in brightness depends on the refractive index of the object.

Applications for Phase Contrast Microscopy

Phase contrast is preferable to bright field microscopy when high magnifications (400 x, 1000 x) are needed, the specimen is colorless or the details are so fine where the color is insignificant. Most of the living microscopic organisms are much more obvious in phase contrast.

Using Phase Contrast

Phase contrast condensers and objective lenses add considerable cost to a microscope, and so phase contrast is often not used in teaching labs except perhaps in classes in the health professions and in some university undergraduate programs.

The light path must be aligned to use phase contrast. An element in the condenser is aligned with an element in a specialized phase contrast lens. This usually involves sliding a component into the light path or rotating a condenser turret. The elements are either lined up in a fixed position or are adjusted by the observer until the phase effect is optimized. Generally, more light is needed for phase contrast than for corresponding bright field viewing, since the technique is based on a diminishment of brightness of most objects.

Microscopy with Oil Immersion

When light passes from a material of one refractive index to material of another, as from glass to air or from air to glass, it bends. This property of light is known as diffraction. Light waves of different wavelengths bends at different angles while passing from one medium to another medium. So when objects are magnified using light waves, the images become less and less distinct. With "dry" objective lenses this loss of resolution prevents using magnifications of above 400x or so. In fact, as you will see later, even at 400 x the images of very small objects are badly distorted. In order to prevent this oil immersion is used.

Oil immersion microscopy is essential to any microbiology lab. Stained smears of mixed culture are recommended for practice.

Principle

Placing a drop of oil with the same refractive index as glass between the cover slip and objective lens eliminates two refractive surfaces, so that magnifications of 1000x or greater can be achieved along with preserving good resolution.

Note : After using oil immersion lenses should be cleaned properly by wiping gently with xylene.

Resolving power : Is the ability to observe two adjacent points as distinct and separate as possible. It is expressed as,

$$\text{Resolving power} = \lambda/\text{NA}$$

NA=numerical aperture, NA=n sin θ, λ-wavelength of the incident light, n=refractive index of the medium (1 for air), θ-angle of incidence of light waves.

Magnification : Usually 10 x is for the eyepiece, and 10 x, 40 x, 100 x (for oil immersion) for objective lense.

Electron microscopy : In electron microscopy a beam of electron is used for the observation of specimen. It gives a magnification up to 400,000 times.

Section 3

BIOCHEMISTRY

Chapter 6

Biochemical Preparations

Preparation of solutions and buffers are the most important part of any experiment in Biotechnology. For accurate preparation of the solutions take care of the following.

1) Weighing should be accurate
2) Solvent selection and concentration should be proper
3) Proper mixing
4) Use of good quality chemicals etc has to be taken care of.
5) After preparation every solution should be labeled and stored.

1. Common solvents

See Appendix D

2. Common Buffer Compositions

Buffer

Acid is a substance which can donate a proton and base is a substance which can accept a proton .So every acid has its conjugate base. Alkali is a substance which can release a hydroxyl group. The ionic substance which can act both as acid and base is known as ampholyte.

In solution an acid or base undergoes ionization and the concentration of H^+ or OH^- ions is a measure of the concentration of acid or alkali. Strong acid and base undergoes complete ionization while weak acid or base dissociates to a lesser extent.

The concentration of H^+ ions can be expressed as $-\log_{10}H^+$ and it is known as pH of the solution.

For water, it is weakly dissociated as $H_2O \rightarrow H^+ + OH^-$

At equilibrium the dissociation constant of water can be given as

$K = [H^+] [OH^-] / [H_2O]$

For strong acids

Strong acids get completely ionized in solution and the hydrogen ion concentration will be same as that of the acid. Then pH can be given as $pH = -\log [H^+]$

For weak acids

Weak acids do not undergo complete ionization so at equilibrium the

$pH = pK_a + \log_{10}[A^-]/[HA]$

Buffer solutions : Have the ability to the resist the change in pH.

Buffer value: The ability to the extent to resist the change in pH

Preparation of Some Common Buffers

pH meter: The instrument used for the measurement of pH of a solution. It consists of a thin glass bulb in the end of a glass rod. Inside the bulb there is a solution of 0.1mol/ litre of HCl connected to a platinum wire. Ag in $AgCl_2$ electrode which is reversible to hydrogen ions. The potential developed in between the thin glass bulb depending on the pH in which it is immersed.

The electromotive force (emf) of the cell can be given as

$$E = E_{ref} - E_{glass}$$

$$E_{glass} = E_{ref} - E$$

New electrode must be kept in distilled water for several hours. It can be kept in 0.1ml HCl or 4M KCl depending on the manufactures instructions. Electrodes must be washed and cleaned thoroughly with distilled water before and after each use.

Standardization : Potassium hydrogen pthalate is the general standard for pH 4. Different standards of pH 4, 7, 9 are available. pH meter should be calibrated with nearest pH with standard pH buffer for accurate measurement. Electrode should be kept in an appropriate solution as per manufactures instruction and should never be dried up.

For common buffer compositions see Appendix B

Chapter 7

Biochemical Estimations

Most of the biochemical estimations using spectrophotometer or calorimeter are based on the Beer-Lambert Law and are depending upon the optical density which is proportional to the concentration of the sample.

1. Determination of λmax

Aim : To determine the **λmax** of the given solution

Principle

When a beam of monochromatic light passes through a colored solution, a part of it get absorbed by the solution and the intensity of the light passes out will be lesser than of the incident light. The absorption depends on the path length and concentration of the solution (Beer-Lambert law). Each colored compound will have a particular wavelength on which the absorption of the light will have a maximum value depending on the atomic bonding.

Materials

Colored solution : 0.1mg/ml in distilled water (for example $CuSO_4$, $K_2Cr_2O_7$, Bromophenol blue etc.)

A colorimeter with different filters required for light waves. When using a colorimeter different wavelength has to be set. In a spectrophotometer scanning option for λ_{max} in different range can be selected.

Procedure

1) Take distilled water in a cuvette as blank.
2) Set the wavelength (220 nm or the lowest wave length) in a colorimeter.
3) Adjust the OD of the blank to zero.
4) Then keep the first sample (0.1mg/ml $K_2Cr_2O_7$) in the cuvette and find out the OD.
5) Set next wavelength [For example 250 nm. As this varies with colorimeter, use the next higher option.]. Adjust blank OD to zero.
6) Keep the sample again and note the OD.
7) Repeat steps 1 to 6 at different wavelength (200nm to 1000nm) and note the OD values. Plot a graph between wavelength on X-axis and OD on Y-axis for each sample.

Similarly note the OD for different samples.

Plot a graph between OD and wavelength for each sample.

Table showing wavelength and OD value

Sample taken	**Concentration**	**Wavelength used (nm)**	**OD value obtained**
$K_2Cr_2O_7$ solution	0.1 mg/ml	220	
		250	
		300	
		350	
		400	
		450	
		500	
		550	
		600	
		650	
		700	
		750	
		800	
		850	
		900	
		950	
		1000	
Copper sulphate	0.01mg/ml	Repeat as above	

The wavelength at which the OD has maximum value is the λ_{max} of that compound.

Results

λ_{max} of the given compound potassium dichromate is ....

Note : If the OD of the sample is not coming in range (i.e. OD value is above 2.0) of the colorimeter, the solution may be diluted. After obtaining the range, narrow the range to one digit and get exact λ_{max} of the compound. For example

Take potassium dichromate solution of concentration 100mg/1000ml and check OD as above. If the maximum value of OD comes in between 200 to 350 nm, again check the OD between the range 250 to 300 nm by 10 digits i.e. at 250,260,270,280,290,300nms.and then between 250 to 260 nm by one digit i.e. at 250, 251, 252, 253, 254, 255, 256, 257, 258, 259, 260nm.

For potassium dichromate, at a concentration of 100mg/1000ml

At 235 nm- OD value is 1.22

At 257nm- OD value is 1.42

At 313nm-OD value is 0.464

At a concentration of 50 mg/1000ml

At 235 nm - OD value is 0.621

At 257 nm - OD value is 0.734

At 313nm- OD value is 0.257

Therefore the λ_{max} of potassium dichromate is 257nm.

2. Verification of Beer-Lambert Law

Aim : To verify the Beer Lambert law.

Principle : When a beam of monochromatic light passes through a colored solution a part of it get absorbed by the solution and the intensity of the light passes out will be lesser than that of the incident light. The absorption depends on the path length and concentration of the solution (Beer-Lambert law). Each colored compound will have a particular wavelength at which its absorption of the light will have a maximum value, depending on the nature of the atomic bond present in it.

If I_0 is the Intensity of the incident light and I is the intensity of the light passes out, then

$I = I_0 C^{-kct}$

k- Proportionality constant

c- Concentration of the absorbing medium.

Extinction coefficient : The expression $\log_{10} I_0/I$ is known as extinction or absorbance.

It is referred to as optical density (OD). A plot of OD against concentration gives a straight line passing through the origin.

Materials

Colorimeter

Colored solution- 100 mg/ml in distilled water (For example: Potassium dichromate, Copper sulphate etc).

Procedure

Pipette out the solution 1ml, 2ml, 3ml, 4ml, and 5 ml serially in separate test tubes.

Make up to 5 ml with distilled water. Mix well.

Take 5 ml of distilled water as blank. Read OD of each solution at λ_{max} (as obtained in experiment no 1).

Plot a graph between concentration on X-axis and OD on Y-axis.

Result

Graph obtained is a straight line passing through origin.

Table: Concentration of the sample against OD

Volume of the solution pip petted (ml)	Volume of distilled water added (ml)	Total volume (ml)	OD value obtained
0.0	5.0(Blank)	5.0	
1.0	4.0	5.0	
2.0	3.0	5.0	
3.0	2.0	5.0	
4.0	1.0	5.0	
5.0	0.0	5.0	

3. Estimation of Carbohydrate by Anthrone Method

Aim: To estimate the quantity of carbohydrate present in the given sample by anthrone method.

Principle: Glucose undergoes dehydration reaction by concentrated acid and as a result furfural and furfural derivatives such as hydroxyl methyl furfural are formed. The concentrated acid added, hydrolyses the glycoside bond to form mono saccharides. It is then dehydrated to furfural and its derivatives. The product combines with anthrone to give a blue green complex with λ_{max}620nm.

Requirements. Test tubes, Pipettes, Test tube stands, Water bath, Colorimeter.

Reagent

1) *Anthrone reagent* : 2gm/l of anthrone in 95% H_2SO_4
2) *Standard* : 100 µg/ml of glucose in distilled water. (working standard 10 µg/ml, i.e. dilute 1ml to 10 ml)

Procedure: Pipette aliquots of 0.0 ml, 0.2ml, 0.4ml, 0.6ml, 0.8 ml and 1.0 ml each of the standard and 0.2 ml and

0.5ml of the unknown samples in to a series of clean dry test tubes.

Make up to 1ml each with distilled water.

Add 4ml of anthrone reagent to each of the test tubes. Mix well.

Incubate for 10 min on boiling water bath.

Cool and take OD at 620nm.

Prepare the standard curve with concentration on the X-axis and OD on the Y-axis.

From this graph the concentration of the given test solution is calculated.

Table 3.2.3: Estimation of carbohydrate by anthrone method.

Volume of standard pipetted (ml)	Volume of distilled water added. (ml)	Final volume (ml)	Concentration of the sample (µg/ml)	Volume of anthrone reagent added (ml)	Incubation time at boiling water bath (min)	OD at 620 nm
0.0	1.0	1.0	0.0	4	10	
0.2	0.8	1.0	2.0	4	10	
0.4	0.6	1.0	4.0	4	10	
0.6	0.4	1.0	6.0	4	10	
0.8	0.2	1.0	8.0	4	10	
1.0	0.0	1.0	10	4	10	
0.2(test)	0.8	1.0	-	4	10	
0.5ml(test)	0.5	1.0	-	4	10	

Result: The concentration of the test solution is

Note: Sulphuric acid should be concentrated and anthrone reagent should be prepared freshly for good results.

The anthrone reagent is very strong and maximum care should be taken in handling.

4. Estimation of reducing sugar by Di Nitro Salicylic (DNS) acid

Aim: To estimate the concentration of reducing sugar by DNS method.

Principle: 3, 5 Dinitrosalicyclic acid (light yellow color) in alkaline solution gets reduced by the sugar to 3-amino,5-nitro salicylic acid which will appear in dark yellow color.

Materials: NaOH-2mol/litre, Na-Ktartarate-30 gm in 500ml water,

DNS-10.0 gm in 200 ml of 2mol/litre NaOH,

DNS reagent: (Mix 500ml of Na-K tartarate and 200ml of DNS in NaOH. Mix well and dilute to 1000ml with distilled water.

Distilled water

Working standard (glucose) - 250µg/ml, calorimeter and boiling water bath.

Procedure: Pipette 0.0 ml, 0.2 ml, 0.5 ml, 0.8 ml, 1.0 ml, 1.5 ml, 2.0 ml, 2.5 ml and 3.0 ml of the standard and 1.0 ml and 1.5 ml of the test solution into a series of test tubes. Make up to 3ml with distilled water. Add 1ml of DNS reagent to each test tube.

Incubate for 5min.on boiling water bath, cool to room temperature.

Take OD at 540nm. Plot a graph between concentration on X-axis and OD on Y-axis. The concentration of the test sample can be calculated from the graph knowing the OD.

Table 3.2.4: Estimation of reducing sugar by DNS method

Volume of the glucose standard taken (ml)	Volume of the distilled water added (ml)	Total volume (ml)	Volume of the reagent added (ml)	Incubation time in Min (on boiling water bath)	OD at 540 nm
0.0	3.0	3.0	1 (blank)	5	
0.2	2.8	3.0	1	5	
0.5	2.5	3.0	1	5	
0.8	1.2	3.0	1	5	
1.0	2.0	3.0	1	5	
1.5	1.5	3.0	1	5	
2.0	1.0	3.0	1	5	
2.5	0.5	3.0	1	5	
3.0	0.0	3.0	1	5	
1.5(unknown)	1.5	3.0	1	5	
0.5(unknown)	2.5	3.0	1	5	

Result : Concentration of the reducing sugar present in the given sample is

5. Estimation of Protein by Biuret Method

Aim: To estimate the quantity of protein in the given sample

Principle: Protein forms a colored complex on reacting with alkaline copper sulphate and its absorption maxima is at 540 nm.

Materials required: Protein standard-5mg/ml in distilled water

1) *Biuret reagent* : Dissolve 3 gm of copper sulphate and 9.0gm of sodium potassium tartarate in 500ml of 0.2mol/litre of NaOH.
2) Water bath at 37^{0}C.

Procedure: Pipette the aliquots of 0.0ml, 0.2ml, 0.4ml, 0.6 ml, 0.8 ml, 1.0 ml, 1.5 ml, 2.0 ml of the standard and

1ml and 1.5 ml of the test solution to a series of test tubes. Make up to 2ml with distilled water. Add 3ml of biuret reagent to each test tube.

Mix well and warm at 37 ^{0}C for 10 min and allow cooling.

Read OD at 540nm.Plot the graph between concentration of standard on X-axis and OD on Y-axis.

Calculate the concentration of the unknown from the graph using the OD value.

6. Estimation of Protein by Lowry's Method

Aim: To estimate the quantity of protein in the given sample

Principle: Protein can react with Folin's Ciocalteau reagent to give a colored complex product .This is due the reaction of protein with copper in alkaline conditions and the reduction of phosphomolybdnate present by tyrosine and tryptophan present in the protein.

Materials required: 1.Protein standard Bovine Serum Albumin (BSA)- 1.0mg/ml in distilled water.

1. Alkaline sodium carbonate solution (20 gm/lit $NaCO_3$ in 0.1mol/litre Na OH)
2. Copper sulphate-sodium potassium tartarate solution (5gm/litre of $CuSO_4$ in 10 gm/ litre of Na, K - tartarate. Prepare fresh.
3. FC reagent-Dilute 1:1 with distilled water.

Procedure: Prepare aliquots of 0.0 ml, 0.01ml, 0.02 ml, 0.04 ml, 0.06 ml, 0.08 ml , 0.1 ml ,0.12ml,0.16ml,0.18ml and 0.2ml of the test solution and 0.1ml and 0.5 ml of the test solution in a series of test tubes. Make up to 1 ml with distilled water. Add 5 ml of the alkaline solution to each test tube. Mix well. Incubate at room temperature for 10min.Add 0.5ml of diluted FC reagent in each test tube. Mix well. Read OD at 750 nm.

Table 3.2.6: Lowry's method for protein estimation

Volume of the protein standard pipette (ml)	Volume of distilled water added (ml)	Total volume (ml)	Volume of alkaline solution added (ml)	Time of incubation at room temperature (min)	Volume of FC reagent added (ml)	OD a 750 nm
0.00	1.0	1.0	5.0	10	0.5	
0.01	0.99	1.0	5.0	10	0.5	
0.02	0.98	1.0	5.0	10	0.5	
0.04	0.96	1.0	5.0	10	0.5	
0.08	0.92	1.0	5.0	10	0.5	
0.10	0.90	1.0	5.0	10	0.5	
0.16	0.84	1.0	5.0	10	0.5	
0.18	0.82	1.0	5.0	10	0.5	
0.20	0.00	1.0	5.0	10	0.5	
0.50	0.5	1.0	5.0	10	0.5	

Prepare a standard curve and estimate the concentration of the test from the graph.

7. Estimation of Protein by Bradford Method

Aim: To estimate the quantity of protein in the given sample by Bradford method

Principle: The assay based on the ability of proteins to bind commassie brilliant blue G 250.The colored complex has an absorption maxima at 595and there is a shift in the OD value of the dye on binding with the protein.

Materials: Commassie brilliant blue G 250-0.1g, dissolve in 50 ml 90% ethanol. Add 100ml of ortho phosphoric acid and make up to 200ml with distilled water. Filter and store in amber colored bottle in a fridge. Dilute 5 times before use with distilled water.

Protein standard- IgG -0.1mg in 100ml PBS

Procedure: Prepare a series of protein concentrations in a series of test tubes. Add 5ml of the dye to each tube. Mix well. Allow to develop color 5 to 10 min. There will be a color change from red to blue on binding with the protein. Read OD at 595 nm. Plot the standard curve and calculate the concentration of the test sample using the graph and OD value of the unknown.

8. Estimation of Cholesterol

Aim: To estimate the quantity of cholesterol in the given sample by $FeCl_3$ method.

Principle: Cholesterol gets reduced in presence of con.H_2SO_4 .This reduced cholesterol reacts with Fe^{3+} in $FeCl_3$ to form colored complex with λ_{max} 560nm.

Requirements: Test tubes, Pipettes, Test tube stands, Colorimeter

Reagent

1) *Ferric chloride*: 840 mg of $FeCl_3$ in 10 ml of glacial acetic acid. From this 1ml is added to 100ml of glacial acetic acid.
2) Concentrated H_2SO_4
3) *Standard*: 1mg/ml of cholesterol in chloroform. Take 10.0 ml of this and again make up to 100ml with chloroform and use as working standard.

Procedure : Aliquots of 0.0ml, 0.2ml. 0.4ml, 0.6ml, 0.8ml, 1.0ml, 1.2ml of the standard and 0.5ml and 1.0 ml of the unknown samples are pipette out to clean dry test tubes kept serially.

Keep at boiling water bath for 10 min.

Add 3ml of $FeCl_3$ to each test tube. Mix well.

Then add 2ml conc. H_2SO_4 in each test tube and mix well.

Incubate for 45 min at R.T. Take OD at 560nm.

Record the observations as follows.

Table 3.2.8: Estimation of cholesterol

Volume of standard pipetted (ml)	Incubation time at boiling water bath (min)	Volume of $FeCl_3$ reagent added. (ml)	Volume of con.H_2SO_4 Added (ml)	Incubation time at room temperature (min)	OD value at 560 nm
0.0	10	3	2	45	
0.2	10	3	2	45	
0.4	10	3	2	45	
0.6	10	3	2	45	
0.8	10	3	2	45	
1.0	10	3	2	45	
1.2	10	3	2	45	
0.5	10	3	2	45	
1.0	10	3	2	45	

Calculate the concentration of the test solution from the graph. The graph can be plotted with concentration of the solution on X-axis and OD on Y-axis

Result: The concentration of the test solution is

9. Estimation of DNA

Aim: To estimate the quantity of DNA in the given sample by Diphenylamine method.

Principle: Deoxyribose of purine nucleotide of DNA reacts under acidic condition to form β-hyroxylevulinyl aldehyde. This condenses with diphenylamine to give a blue colored complex with absorption maxima at 595nm.In DNA only deoxyribose of the purine nucleotide reacts to give this reaction. So the value obtained represents theoretically half of the total deoxyribose present.

Requirements: Test tubes, Pipettes, Test tube stands, Colorimeter.

Reagent

1) *Buffered saline*: Sodium citrate buffer 150mM, pH-7, NaCl-0.15mM, Na citrate-0.015mM
2) *Diphenylamine reagent*: Dissolve 1 gm of diphenylamine reagent in 100 ml of glacial acetic acid followed by an addition of 2.5 ml of conc. H_2SO_4.Mix well and prepare half an hour before and keep in amber colored bottle.
3) *Standard*: 50 to 500μg/ml. (prepare 100 μg /ml in buffer (reagent no.1 as standard)

Procedure: Pipette aliquots of 0.0ml, 0.2ml. 0.4ml, 0.6ml, 0.8ml, 1.0ml.1.2ml, of the standard and 1.0ml, 2.0ml of the unknown samples to clean dry test tubes which are kept serially.

Make up to 2ml with distilled water.

Add 4ml of diphenylamine reagent. Mix well.

Incubate for 10-15 min on boiling water bath.

Cool and take O.D at 595nm. Plot the graph. Calculate the quantity of DNA in the test solution.

Table 3.2.9: Estimation of DNA

Volume of standard DNA solution pipetted (ml)	Volume of distilled water added (ml)	Total volume (ml)	Volume of di phenyl amine added (ml)	Incubation time on boiling water bath (min)	OD value at 595 nm
0.0	2.0	2.0	4.0	15	
0.2	1.8	2.0	4.0	15	
0.4	1.6	2.0	4.0	15	
0.6	1.4	2.0	4.0	15	
0.8	1.2	2.0	4.0	15	
1.0	1.0	2.0	4.0	15	
1.2	0.8	2.0	4.0	15	
1.0	1.0	2.0	4.0	15	
2.0	0.0	2.0	4.0	15	

Find out the concentration of given test sample from the standard curve plotted.

Result: The concentration of the test solution is:

Note: Sensitivity-50-500 µg/ml.

10. Estimation of RNA

Aim: To estimate the quantity of Ribo Nucleic Acid (RNA) in the given sample by Orcinol method.

Principle: Orcinol reaction is a general reaction for pentose. The acid in the orcinol reagent cleaves some of the phosphodiester bonds and hydrolyses glycosidic link between ribose and purines. The hot acid also converts ribose to furfural. Furfural gives many reactions analogous to benzaldehyde and it condenses with orcinol to give a green colored condensation product triacyl methane derivative. The color formed depends on the concentration of HCl, $FeCl_3$, Orcinol, Heating time etc.

Requirements: Test tubes, Pipettes, Test tube stands, Water bath, Colorimeter

Reagents

1) *Buffered saline*: Sodium citrate buffer: 150mM, pH-7, NaCl-0.15mM, Naiscitrate-0.015mM
2) *Orcinol reagent* : Dissolve 1.2 gm of orcinol in 20 ml of ethanol.
3) 0.5 $FeCl_3.6\ H\ _2O$ in 500ml of conc. HCl
4) Mix 17.5 ml of orcinol +500 ml of $FeCl_3$

Standard: 0.2mg/ml of RNA in buffer saline (reagent no.1)

Procedure : Pipette the standard nucleic acid of volume 0.0 ml, 0.2ml, 0.4ml, 0.6ml, 0.8ml, 1.0ml, 1.5ml, 2.0ml, 3ml and 1ml and 2ml of test/unknown solution in a series of tubes.

Make up the volume to 3ml in each tube with distilled water.

Add 3ml of orcinol reagent. Mix well.

Incubate for 20 min on boiling water bath. Cool and take OD at 665nm.

Plot the graph between concentration of standard on X-axis and OD on Y-axis.

Find out the concentration of given test sample form the standard plot using the observed OD of the test sample.

Result: The concentration of the test solution is

Note: The yield of the RNA isolated can be estimated by using it as a test solution.

Table 3.2.10: Estimation of RNA

Volume of standard RNA solution pipetted (ml)	Volume of distilled water pipetted (ml)	Total volume (ml)	Volume of orcinol reagent pipetted (ml)	Incubation time on boiling water bath (min)	OD at 665nm
0.0	3.0	3.0	3.0	20	
0.2	2.8	3.0	3.0	20	
0.4	2.6	3.0	3.0	20	
0.6	2.4	3.0	3.0	20	
0.8	2.2	3.0	3.0	20	
1.0	1.0	3.0	3.0	20	
1.5	1.5	3.0	3.0	20	
2.0	1.0	3.0	3.0	20	
3.0	0.0	3.0	3.0	20	
1.0(test sample)	2.0	3.0	3.0	20	
2.0	1.0	3.0	3.0	20	

11. Determination of ISO Electric Point of Amino Acid

Aim: To determine the ISO electric point of the given amino acid by preparing the titration curve.

Principle: At neutral pH amino acids are present as zwitter ions and are amphoteric in nature. It can be titrated against both acid and alkali. This is due to the presence of -COOH and -OH group.

Materials

HCl: 0.1 N-100ml

NaOH: 0.1 N-100ml

Amino acid (example: Glycine: 0.1 N-100ml)

pH meter

Procedure: Standardize the pH meter. Pipette 10ml of the amino acid in to a clean dry beaker. Check the pH of the amino acid. Add 5ml of the 0.1 N HCl to the beaker. Mix gently. Check pH. Note the value. Similarly keep on adding HCl till the pH lowers by one digit. Then add the acid in greater volumes till the pH reaches 1.5.Note the pH value against the volume of the acid added.

Take another 10 ml of the 0.1N glycine in a beaker. Note the pH. Add 0.1N NaOH little by little till the pH increases by one digit. Add NaOH in more volume till the pH reaches 12.0.Note the volume added against pH obtained each time. Plot a graph between volumes of the acid/alkali added on X-axis and pH on Y-axis.

Calculate the isoelectric point using pK_a and pK_b from the graph and compare with the standard value.

Chapter 8

Biochemical Separations

1. Chromatography

Theory : Molecules can be separated from each other on the basis of their size shape, mass and charge and adsorption properties. The Russian botanist Twestt used this technique for the first time for the separation of colored plant pigments on alumina.

The major types of chromatography are

I) *Gel filtration*: Depending on the size of the particle it will be eluted out. i.e. smaller particles will be eluted out first.

II) *Adsorption chromatography*: Depending on the adsorption property of the components it will get adsorbed on the inert material and less adsorbed component will be eluted out first.

III) *Ion exchange chromatography*: Cation and anion exchangers are used as the matrix of this chromatography. Depending on the charge of the particle it will get bonded on the matrix and other will be eluted out. The bonded particle can be eluted out using a suitable buffer.

IV) *Partition chromatography*: Depending on the solubility of components in a mixture of solvents the separation can be done.

Separation of Amino Acids by Paper Chromatography

Aim: To separate the amino acids in the given sample.

Principle: The components in a mixture can be separated on the basis of their partition coefficient in two immiscible phases. The stationery phase may be solid or liquid while mobile phase is a liquid or gas. Depending on the stationery phase the chromatography can be

1) Paper Chromatography
2) Thin Layer Chromatography
3) Ion exchange Chromatography
4) High Pressure Liquid Chromatography

Paper Chromatography

Separation is based on the liquid-liquid partition of amino acids between the water molecules adsorbed on the cellulose layer of the paper and the mobile organic phase.

Materials Required

1) Whatman No.1 sheets
2) Solvent system (n-butanol: acetic acid: water -12:3:5)- Prepare minimum 3 hrs prior the running)

3) Detecting reagent-Ninhydrin-0.2%in acetone-100ml-prepare fresh
4) *Amino acid standards*: 1mg/ml (Example: Aspartic acid, Serine, Glycine, Alanine, Proline, Arginine, Lysine, Histidine)
5) Chromatographic chamber
6) Sprayer
8) Dryer

Procedure

Cut the What man no 1 sheet into pieces of 30 x 15cm size.

Draw a line at a distance of 5 cm from bottom.

Mark the points at equal distances on the line for loading the samples.

Draw a line at 3 cms from the top end also.

Load each amino acid standards (5 micro liters each) on equi-distant points on the bottom line. Load the mixture of amino acids or an unknown sample on the last point. Dry it.

Put the paper in the saturated chamber and allow to run till the solvent front reaches the top line. Take the paper out, dry it.

Then spray the ninhydrin reagent.

Develop the paper by keeping in a hot air oven at 105^0C for 5 min.

Spots will be developed and measure the distance moved by the solvent and each amino acids.

Results

R_f value of the given amino acid= Distance moved by the amino acid/Distance Moved by the solvent front

Rf value of the given amino acid

$$= \frac{\text{Distance moved by the amino acid}}{\text{Distance moved by the solvent front}}$$

The amino acids of the given mixture got separated into individual amino acids and identified with the R_f values.

Note : For TLC documentation formate see 'Appendix F'.

2. Identification of Sugars from Fruit Juice by Thin Layer Chromatography

Aim : To identify the sugars present in the given fruit juice.

Material Required

1) *Silica gel plates*: Prepare the plates by mixing Silica gel (TLC grade) with distilled water in the 1:2 ratio. Make 100 ml of uniform slurry
2) *Solvent system*: Prepare the solvent system by mixing the solvents Pyridine, Ethyl acetate, Acetic acid, water in the ratio 5:10:3:1 (Prepare 400 ml)
3) *Spray reagent*: Aniline diphenylamine reagent-Mix equal volumes of 1% Aniline and 1% di phenyl amine in acetone. Then it is mixed with 1 volume of 85% phosphoric acid i.e. in the ratio 5:5:1
4) *Sugar standards*: 1% solutions of glucose, fructose, galactose, mannose in 10% isopropanol.
5) Fresh fruit juices- (Lemon, Tomato, Orange etc)-1 ml each (Add 3ml of ethanol to 1ml of juice to precipitate denatured proteins).
6) TLC SET

Procedure

Prepare silica gel plates by mixing silica gel with distilled water in the ratio 1:2. (Add a pinch of CaSO4 as a binder if required).

Make uniform slurry and fill in the filling chamber.

Place it at one end of the plates and drag it over the plates carefully to get a uniform thickness. Dry it.

Pre-heat the plate at 110^0 C to get activated.

Load the samples on the silica gel plate by leaving 2cms from the bottom. Different samples/standards (10 micro liters each) are to be loaded at equal distance in separate points Dry it.

Keep it in the equilibrated TLC tank.

Allow to run till the solvent front reaches 2cms away from the top. Take out. Dry it.

Spray the detecting agent. Dry. Develop it at 105^0C in a hot air oven for 20 min.

Results

Points were detected and compared with the standards. Calculate R_f values for each spot. Compare with the standards.

Note: For more TLC solvent system and detection see Section 6.2.5.

3. SDS-PAGE (Poly Acryl amide Gel Electrophoresis)

Aim: To separate proteins according to their size by SDS PAGE.

Principle: Electrophoresis can be used to separate and identify proteins by applying electric current. Electrophoresis is done using poly acryl amide gels with SDS.

SDS

In order to separate different protein molecules of different shapes and sizes, it is necessary to denature the protein so that the proteins no longer have any secondary, tertiary or quaternary structure. Consider two proteins each of 500 amino acids long. Let one is shaped like a closed umbrella while the other one looks like an open umbrella. The 3D structure of a molecule will determine how well it can move through an environment. SDS can be used to denature all proteins to the same linear shape. SDS (sodium dodecyl sulfate) is a detergent that can dissolve hydrophobic molecules but also has a negative charge (sulfate) attached to it.

Therefore, if a cell is incubated with SDS

1. The membranes will be dissolved,
2. All the proteins will become soluble by the detergent,
3. All the proteins will be covered with many negative charges and
4. All proteins have a large negative charge which means they will all migrate towards the positive pole when placed in an electric field.

PAGE

If the proteins are denatured and put into an electric field, they will all move towards the positive pole at the same rate, with no separation by size. So we need to put the proteins into an environment that will allow different sized proteins to move at different rates. The environment of choice is poly acryl amide, which is a polymer of acryl amide monomers. When this polymer is formed, it turns into a gel and we will use electricity to pull the proteins through the gel so the entire process is called polyacrylamide gel electrophoresis (PAGE). A polyacrylamide gel is not solid but is made of layers of tunnels through a meshwork of fibers.

If all the proteins enter the gel at the same time and have the same force pulling them towards the other end, because of their small size, smaller proteins will move through faster than the larger one. The mixture of proteins of given size tend to move through the gel at the same rate, even if they do not take exactly the same tunnels to get through. Likewise, proteins tend to move through a gel in bunches, or bands, since there are so many copies of each protein and they are all the same shape and size. So after the electrophoresis proteins bands of different size will be visible on the gel, i.e. separated according to their molecular size as smaller one at the bottom of the gel.

NB: Acrylamide monomer is a potent cumulative neurotoxin. Do not mouth pipette acrylamide solutions, and wear gloves when handling unpolymerised solutions.

Assembling Gel Apparatus

Assemble two glass plates (one notched) with two side spacers, clamps, grease, etc. as shown by demonstrators. Stand assembly upright using clamps as supports, on glass plate. Pour some **pre-heated 1% agarose** onto glass plate, place assembly in pool of agarose; this seals the bottom of the assembly.

Resolving Gels

Gel concentration of 12.5% in 0.25 M Tris-HCl pH 8.8

	Volume Reagent: (to make 30 ml)	**Volume (to make 10 ml)**
40% Acryl amide stock*:	9.4	3.1
Water (distilled)	12.3	3.8

1 M Tris-HCl pH 8.8	7.5	2.5
10% SDS	0.3	0.1
Ammonium per sulphate 1% (polymerizing agent)	0.5	0.5
TEMED (added last)	20.0 μl	20.0 μl

* = 19:1 - 38:1 w: w ratio of acryl amide to N, N'-methylene bis - acryl amide

Mix ingredients gently in the order shown above, ensuring no air bubbles form. Pour into glass plate assembly. Overlay gel carefully with isopropanol to ensure a flat surface and to exclude air. Wash off isopropanol with water after gel has set (±15 min).

Stacking Gels

Gel concentration of 4.5% in 0.125 M Tris-HCl pH 6.8

	Volume Reagent: (to make 15 mL) ml	**Volume (to make 10 ml) ml**
40% Acryl amide stock	1.7	1.1
Water	10.8	7.1
1 M Tris-HCl pH 6.8	1.9	1.25
10% SDS	0.15	0.1
Ammonium per sulphate 1%	0.5	0.5
TEMED (stir quickly)	20 μl	20 μl

Mix as before, then **pour onto top of set resolving gel,** insert comb, allow to set the gel and then remove the comb. Assemble top tank onto glass plate assembly. Fill with electrophoresis buffer.

Electrophoresis buffer

The final tank buffer composition is 196mM glycine / 0.1% SDS / 50mM Tris-HCl pH 8.3, made by diluting a **10 x stock** solution. This goes in both top and bottom tanks.

Samples

Grind a little leaf material (e.g. 2 grams) in a mortar in the extraction buffer. Centrifuge in an eppendorf tube for 3 min. Take supernatant and mix it with 100ul of disruption mixture in the ratio 1:1 (this is 125mM Tris-HCl pH 6.8 / 10% 2-mercaptoethanol /10% SDS).For liquid/purified samples: Take sample 100 µl of sample and add 50 - 100 µl of disruption mix. Composition of the extraction buffer varies according to the material from which the protein is extracted. For more details see section 4.2.

Heat the sample in eppendorf's for 5 min at 95 ^{0}C in a "float rack" in a water bath. Load the samples in the wells. Connect the apparatus to power supply and do electrophoresis.

Staining of Gels

Coomassie Brilliant Blue (CBB)/Page-Blue 83: 0.2% CBB in 100ml of the mixture of methanol: water: acetic acid in the ratio 45:45:10 %.Take the gel in a plastic box or glass tray and cover it with staining solution, seal the plastic box and leave overnight on shaker (RT) or for 2 to 3 hours at 37^0 C with agitation. De-stain with 25%: 65%: 10% methanol: water: acetic acid mix, with agitation. Observe in gel doc with protein filters.

Note : Be careful with the reagents of SDS PAGE as some of them are neurotoxin and is poisonous. Chemicals should be of good quality for best result.

4. Composition for Native PAGE

Reservoir buffer : Tris glycine-pH-8.3

3g of tris buffer + 14.4g of glycine. Dissolve in distilled water. Adjust pH with HCl and final volume make up to 1000ml.

Staking Gel

0.5 M of tris base-pH 6.8

6gm of tris base in 40 ml of water. Adjust pH to 6.8and make up to 100ml.Filter and store at 4^0c.

Resolving Gel

3M Tris-pH-8.8,

36.3 gm of tris in 48 ml distilled water, Dissolve. Adjust pH to 8.8 with 1M HCl. Filter and store in fridge.

Acryl amide: Bis acryl amide = 30:08

Stacking Gel

Acryl amide, bis acryl amide-2.5 ml, tris buffer pH6.8-5ml, Ammonium per sulphate-1.5%-1.5ml, Distilled water-11ml, TEMED-100µl.

Separating Gel

Acryl amide, bis acryl amide-7.5 ml,

tris buffer pH8.8-3.75ml,

Ammonium per sulphate-1.5%-1.5ml,

Distilled water-19.75ml,

TEMED-100µl

5. Silver staining of protein gels

Silver staining is more sensitive and even 0.1µg of protein will be visible.

Aim: To silver stain the protein gel.

Principle : The amino acids in the protein reduce silver nitrate to form a visible brown colored complex.

Materials : Protein gel,

Silver nitrate-0.1%

Solution 1-0.1 gm of silver nitrate in 5 ml distilled water. Make 50 ml of 0.36% Na OH and 5ml of 0.1% ammonium hydroxide. Solution 2 - Mix 20 ml of Na OH and 5ml of Ammonium hydroxide. Mix solution 1 and 2.Mix well. Make up to 100ml with distilled water.

Sodium thio sulphate solution- 0.1 %in distilled water.

Developer: Dissolve 3gm of sodium carbonate in 80ml of water. Add 1ml of 1%sodium thio sulphate and 1ml of 37% formaldehyde and make up to 100 ml with distilled water.

Stopper : 5% acetic acid

Procedure: Wash protein gel with distilled water with shaking. Soak the gel in silver nitrate solution for 10 min with shaking. Wash the gel with water. Place the gel in developing tray and pour the developer. Shake gently for some time. Dark brown bands will appear. Stop the reaction by adding 5% acetic acid.

Note: Enough volume of silver nitrate, developer solution etc is to be added to the developing tray to soak the gel.

Section 4

ENZYMES

Chapter 9

Enzymes

Enzymes are biocatalysts composed of amino acids. Enzyme reactions are highly specific. It binds to the substrate to form a reversible intermediate product called enzyme substrate complex. As the concentration of the substrate increases the rate of formation of the intermediate complex also increases. This increase in velocity of the reaction is in a hyperbolic manner till it achieves the maximum velocity called Vmax .At this point all the enzymes will be in the form of ES complex. Substrate concentration will be in at its optimum level i.e. further increase in substrate concentration will not affect the rate of the reaction. The substrate concentration at which the rate of the reaction is half the maximum velocity Vmax is called the Michaelis-Menten constant, Km. This is constant for the given enzyme acting on the same substrate.

This relationship can be given by the formula

$$V_0 = V_{max} [S]/K_M + [S]$$

V_0 is the initial rate at the substrate concentration [s].

1.0 unit of enzyme activity is defined as the amount causing the transformation of 1.0 micromole of the substrate per minute at 25° C under the optimal conditions. Specific activity of the enzyme is the number of the enzyme units per milligram of protein.

All enzyme reactions have optimum pH and optimum temperature of reactions. Enzymes can be inactivated by the modifications of functional group present and essential for the reactions. Inactivation also occurs by irreversible binding of the functional group of the enzymes by competitive and non competitive inhibitors. Competitive inhibitors have structural similarity with the substrate. So they can competitively bind with the enzyme instead of the substrate and thus inhibits the reaction. Non competitive inhibitors bind to some other site on the enzyme and alter the enzyme structure. This causes the inhibition of the enzyme reactions.

In addition to the catalytic activity some enzymes possess regulatory activity and serve as the control of the rate of the metabolism. These are called allosteric enzymes. The reaction rate is controlled by the reversible non covalent binding of the regulator or modulator to the modulator site on the enzyme. This modulator can be substrate or any other metabolic intermediates.

Some enzymes occur in multiple forms called isozymes. The kinetic characteristic will be different for isozymes.

Enzyme Extraction

Enzymes can be extracted from different sources. The procedure for enzyme extraction varies from enzyme to enzyme. It also depends upon the source material from which the enzyme is extracted. Generally for enzyme extraction, collect the material (For example, young plant leaves, germinating seeds, mice liver, microbial culture etc.).Wash it in distilled water. Maintain cold conditions to avoid denaturing of the enzymes. Weigh 2 to 5 gm of the material. Crush it using a mortar and pestle (sterile) in the buffer at pH 7.2. Buffer, mortar and pestle etc should be kept in ice for 1 hr prior to use. Centrifuge at 8000 rpm for 15 min., collect the supernatant. This can be used as a crude extract. It can be further purified using ammonium sulphate precipitation and dialysis. Exact method vary from enzyme to enzyme (Refer section 4.2).

Chapter 10

Enzyme Assays

1. Assay for Catalase

Hydrogen peroxide: Hydrogen peroxide oxido reductase EC1.11.1.6

The decomposition of hydrogen peroxide is catalyzed by the enzyme catalase to give water and oxygen.

Principle: Hydrogen peroxide has absorption maxima at 240 nm. When catalase is added to hydrogen peroxide it decomposes to give water and oxygen and this causes a decrease in absorption with time which can be detected spectrophotometrically.

Materials: Phosphate buffer-0.1 M, pH - 7.2, H_2O_2-10% in phosphate buffer, Enzyme extract-Collect 1 to 2 gram of plant tissue. Clean and rinse the plant material with distilled water. Homogenize in a warring blender with 10 ml of phosphate buffer at 4° C and centrifuge at 12,000 rpm for 20 minutes. Collect the supernatant and use for the assay.

Control: Enzyme solution with phosphate buffer.

Procedure

3 ml each of H_2O_2 in phosphate buffer are taken in 5 test tubes.

Add 0.01 ml, 0.02 ml, 0.03 ml, 0.04 ml, 0.05 ml of enzyme solution to each test tube respectively.

Mix well and immediately place in spectrophotometer and take OD at 240 nm.

The time to decrease the OD by a factor 0.5 is noted. This is taken as Δt.

Calculations: Volume of enzyme extract/Δt = units of enzyme in the reaction mixture.

Enzyme activity -=Decrease in absorbance per minute.

2. Assay for Peroxidase (Oxido Reductase EC.1.11.1.7)

Peroxidase catalyses wide range of reactions like dehydrogenation of organic compounds such as phenols, hydroquinone etc. Eg: Horse radish peroxidase.

o-dianisidine can be used as a substrate for peroxidase. On adding the extract to a solution of o-dianisidine, the reaction starts and the rate of the reaction can be measured by decrease in absorbance at 430 nm

Materials: Phosphate bffer-0.1 M (pH-7.0), o-dianisidine solution-1mg/ml in methanol

Procedure: Take 3.5 ml of buffer in a test tube.

Add 0.2 ml extract and then 0.1 ml of freshly prepared substrate solution.

Bring the mixture to 30° C and add 0.2 ml of H_2O_2. Mix and immediately start reading in a spectrophotometer.

Read initial absorbance and OD at 30 sec intervals for 5 min. The dilution of the enzyme should be proper for good result. If the reaction rate is too high dilute the enzyme extract with buffer. Plot the graph between absorbance and time. Calculate the change in absorbance per min.

Enzyme activity =rate of decrease in absorbance per unit time per 1mg of tissue weight.

Distilled water can be used as blank.

3. Assay for Poly Phenol Oxidase (Mono Phenol, Di Hydroxy Phenyl Alanine: Oxygen Oxido Reductase EC 1.14.18.1)

Phenol oxidase is an important enzyme in plant phenolics product formation. It catalyses the reaction of phenolic compounds to quinones.

The enzyme can be assayed using catechol as substrate. On adding enzyme to the reaction mixture the decrease in OD at 495 nm is to be measured.

Materials : Phosphate buffer: 0.2 M (pH6.5), catechol-0.01 M, Enzyme extract-2 to 5 gm of leaves grind in a pestle and mort or with the 50 ml of Tris HCl buffer (50mM pH 7.2,0.2 M sorbitol,5mMNaCl.)Centrifuge at 10,000 rpm for 15 min. Collect the supernatant for the assay.

Procedure : Take 0.3 ml of catechol solution in a cuvette.

Add 3 ml of the 0.1 phosphate buffer to it and then 0.2ml of the enzyme extract is added and mix well.

Take initial OD. Absorbance (I) is measured at 30 sec intervals for 5 min.

Calculation : Change in absorbance per min=

Unit of enzyme in the test =k x (ΔA/min)

K=0.272 for catechol.

4. Assay for Phosphatases

(Ortho phosphoric mono ester phosphatase, alkaline medium 3.1.1.1)

Phosphatase liberates inorganic phosphate from organic phosphate esters. Alkaline phosphatase can hydrolysis the phosphate esters.

Principle: The enzyme hydrolysis the substrate-p nitro phenol phosphate to p-nitro phenol and inorganic phosphate (pi). The product p-nitro phenol is yellow in color in alkaline medium and absorption maxima at 405nm.The pH optimum is 5.4 for acid phosphatase and 10.5 for alkaline phosphatase.

Materials: Substrate: p-nitro phenol phosphate-0.025gm in 100ml distilled water. Add 1.49gm of EDTA and 0.84 gm of citric acid. Mix well.

Standard: p-nitro phenol solution-100mM in distilled water (fresh).

Na OH -0.1 N in distilled water.

Enzyme extract: Homogenize 1 gm of fresh cascutta filaments in ice cold10 ml of 40 mM citrate buffer pH5.3in a pre chilled mortar and pestle. Filter through 3 layered muslin cloth. Centrifuge at 10,000 rpm for 10 min. Collect supernatant and use as enzyme extract.

Procedure: Incubate 3 ml of the substrate at 37 0C for 5 min in a test tube.

Add 0.5 ml of the enzyme extract. Mix well.

Take 0.05 ml of this the mixture and add 9.5 ml of the Na OH solution.

Take OD at 405 nm. This is the blank. Incubate the remaining enzyme + substrate for 15 min at 37° C.

Take 0.5 ml of the solution form this and add 9.5 ml of Na OH to stop the reaction.

Take OD at 405 nm.

Take 0.2ml, 0.4ml, 0.6ml, 0.8ml and 1.0 ml each of the standards and dilute each to 10 ml with Na OH. Take OD. Draw the standard curve.

Result : The specific activity can be calculated as millimoles of p-nitro phenol released per min per gram of the tissue.

Note : The assay can be repeated for alkaline phosphatase with the following modifications

1. Buffer-10mM glycine, NaOH buffer-pH10.5.
2. *Substrate solution* : 375 mg glycine, 10mg $MgCl_2$, 165mg p-nitro phenyl phosphate. Dissolve in 0.1 NaOH and make up to 100ml with distilled water. Adjust pH to 10.5.

Chapter 11

Enzyme Extraction

1. Extraction of β-galactosidase from yeast cells.

Aim : To extract β-galactosidase from yeast.

Principle : β-galactosidase catalyses the hydrolysis of lactose into its monosaccharide unit of glucose and galactose. Allolactose is also produced by this reaction and the quantity of this controls the expression of the enzyme at the level of transcription. The gene for β-galactosidase synthesis is a set of gene with regulatory gene sequence and it is collectively known as lac operon. It consists of a promoter sequence, an operon, gene for β-galactosidase synthesis lac z and lac y which encodes for permease, a transmembrane protein that allows the transport of lactose entry to the cell. Lac A encodes for thiogalactoside transacetylase, an enzyme whose function is not fully understood. The promoter sequence within the operon is the site where the RNA polymerase binds the DNA and initiates transcription of lac Z, lac Y and lac A genes. The operon sequence that adjacent to the promoter is an important regulatory sequence that will participate in the control of expression of lac operon. Transcription of lac operon can be induced in the presence of lactose and repressed in the absence of lactose. When components required for the metabolism of lactose are not needed i.e. when lactose is absent, lac operon is repressed. ONPG (O-nitro phenyl β-D galactopyran) is used as a substrate for β-galactosidase assay. β-galactosidase will hydrolyze the ONPG substrate to produce two products, one of which absorb at 420 nm (o-nitro phenol) under alkaline conditions.

Requirements: Yeast granules, filter paper, Mortar and pestle.

Lactose broth : Lactose-1gm, Ammonium phospshate-5 gm, Potassium phosphate-1 gm, NaCl-5gm, Magenusium-0.2 gm, Distilled water-1000ml.

ONPG-4mg/ml in distilled water.

Phosphate buffer-1M potassium di hydrogen phosphate 100ml, 1M di potassium hydrogen phosphate 100ml pH-7.

Procedure: Prepare lactose broth and sterilize it.

Take 20 gm fresh yeast granules to 500ml of lactose broth and kept for incubation for 24 hrs at room temperature.

After incubation centrifuge the culture at 5000rpm for 30 min and take the pellet of yeast cells. Crush it with broken glass pieces and centrifuge again for 20 min at 10000 rpm.

Collect the supernatant. This can be used as enzyme extract.

Take 3 ml of buffer and add 2ml of ONPG solution in a test tube. Repeat this in another test tube also.

Add 1ml of enzyme extract to the above mixture in one of the above test tubes.

One test tube without enzyme extract serves as blank. Incubate at room temperature. Note the color change for enzyme reaction.

Result : No color change in the blank. Yellow color will be observed in the test solution which indicates the β-galactosidase activity.

2. Measurement of Enzyme Activity

Activity of β-galactosidase can be estimated using the substrate ONPG.

The enzyme activity can be measured using the rate of disappearance of substrate or the rate of appearance of products with time

$$A \rightarrow P$$

Velocity of the reaction, $V = -d[A]/dt = d[P]/dt$

A graph can be plot by taking amount of substrate against the time. A progress curve will be obtained. The enzyme activity can be obtained by linear part of the curve.

Materials: Phosphate buffer, pH-7 (1.2 gm of K_2HPO_4 and 0.34 gm of KH_2PO_4. Make up to 1000ml with distilled water).

ONPG-4mg/ml,

Crude enzyme extract of previous experiment,

Spectrophotometer and cuvettes.

Procedure : 1 Yeast granules were added to the lactose broth medium and incubate for 24 hrs at room temperature.

Centrifuge at 8000 rpm for 20 min.

Crush the yeast pellets in phosphate buffer & centrifuge again at 8000 rpm for 20 minutes. Collect the supernatant. This can be used as crude extract for enzyme assay.

Pipette 4ml of phosphate buffer, add 0.5ml ONPG. Make up to 5ml with buffer. This serves as control.

In the second test tube take 4ml of phosphate buffer, 0.5 ml of ONPG, and 0.5 ml enzyme extract. Mix well. Immediately record OD at 420nm by using spectrophotometer at an interval of 1 min. till get constant reading. Plot the graph between absorbance against time.

The activity of the enzyme can be calculated according to the formula

$$\text{Activity of the enzyme} = \frac{(\Delta A420/\ \Delta T) \times (1.25/\Delta A420\ max)}{V_P}$$

V_P-volume of the enzyme extract used, ΔT-time interval, ΔA_{420} absorbance at 420

The enzyme activity obtained=----micromole/min/ml

3. Immobilization of Enzymes

Aim : To immobilize the β-galactosidase enzyme using sodium alginate by entrapment method.

Principle : The linkage of enzyme to an inert ligand on an insoluble matrix either by adsorption or by binding will help in immobilizing the enzyme. It will retain its enzyme activity even after the binding with the ligand. When enzyme is added to melted sodium alginate and dropped to $CaCl_2$ solution, calcium alginate will form beads with enzyme entrapped in it.

Materials

Sodium alginate: 4% in distilled water, Melt it at 55° C in a water bath. Keep it in warm conditions to avoid solidification.

$CaCl_2$ solution-0.2 M

Crude enzyme extract

Magnetic stirrer, Pipettes, ONPG (Ortho nitro phenyl β D- galactoside) -4mg/ml

Procedure: Mix 1 volume of enzyme extract with 2 volumes of melted sodium alginate at 50° C. Mix well.

Add this drop by drop to 0.2 M $CaCl_2$ solution kept on a magnetic stirrer.

Collect good beads.

Put the beads in an ONPG solution and incubate overnight.

Note the color change.

Result : Color change is an indication of enzyme activity.

4. Precipitation of Enzymes and Proteins

Aim : To precipitate the crude protein or enzyme by ammonium sulphate precipitation.

Principle : Proteins can be precipitated by salting out method. Usually ammonium sulphate is used for the precipitation. Ammonium sulphate is highly soluble in water and gives high ionic strength. As the concentration of ammonium sulphate increases the solubility of proteins decreases and it starts precipitating out. This is called salting out of proteins. Because of the differences in amino acid content and structure of proteins the salting out of proteins also varies. So proteins can be fractionally precipitated out.

Materials required: Ammonium sulphate -Make it powder. Weigh according to the table below to get 20%, 40%, 50%, 70%, 90% and 100% saturation. Beaker, magnetic stirrer, beads, distilled water, Enzyme extract in phosphate buffer.

Procedure: Prepare an overnight grown yeast culture. Centrifuge at 6000rpm for 30 minutes. Collect the pellet. Grind the pellet in a mortar and pestle. Wash it in sterile phosphate buffer. Centrifuge again. Collect the supernatant and make up to the volume of 50 ml with phosphate buffer. Mix well. Store the enzyme extract in cold conditions. This can be used as enzyme extract for further analysis.

Precipitation of Enzyme: Take 10 ml of the enzyme in a beaker in cold conditions.

Add ammonium sulphate with constant stirring, required for 20% saturation Mix well. Centrifuge and collect the pellet.

The supernatant can be used for further precipitation with different saturation of ammonium sulphate.

Ethanol also can be used for the precipitation of the total protein. The precipitate can be tested for protein by Lowry's method.

Observation: The quantity of protein obtained is

5. Protein Purification by Dialysis

Aim : To purify the isolated protein by dialysis.

Principle : Dialysis can be done with dialysis bag or tubing which are commercially available as dialysis membrane. It selectively permits the passage of solute molecules. So when a mixture of molecules are kept in dialysis bag and put in its solvent (usually buffer or distilled water) the membrane allows the passage of only molecule to the solvent through it and the other will be retained inside the dialysis bag. Due to reverse osmosis ions will migrate from its region of higher concentration to its region of lower concentration i.e. from the bag to the solvent in the beaker. Equilibrium will be reached by the passage of the components to the solvent. So occasional changing of the solvent in the beaker is necessary for the complete purification. Depending upon the pore size of the membrane the component of our interest can be retained inside the dialysis bag. So by selecting a suitable membrane, the component can be purified.

Materials: Dialysis membrane, Buffer, clips, magnetic stirrer, protein mixture to be purified etc.

Processing of dialysis membrane: Cut the given dialysis membrane to a suitable size.

Boil with 1% Na bisulphite in 1mM EDTA for 10 min in a beaker.

Wash with distilled water and again boil in 1mMEDTA for 10 min.

Put in 0.2mM EDTA or 50% alcohol and store in fridge.

The dialysis membrane should not be dried up.

Separation of Protein

The protein/enzyme precipitated by ammonium sulphate will contain ammonium sulphate and it has to be removed. For removing this ammonium sulphate, dialysis can be done.

Procedure: Collect the protein mixture. Take the mixture in a processed dialysis bag closed at one end. Make sure that it is not leaking, before filling with protein mixture. Tie the other end also tightly with dialysis clip or twine. Put the dialysis bag in 2 liters of phosphate buffer kept in a beaker. Keep it on the magnetic stirrer in a cold room for 24 hrs. Rotate in small rpm for mixing of the solvent. Change the phosphate buffer occasionally.

Since ammonium sulphate is smaller in size, it passes through the membrane and protein being larger in size can not pass through the membrane and retain inside the bag. The dialysis membrane of proper size should be selected.

Note: The dialysis membrane of suitable pore size has to be selected. Selection of pore size of the membrane depends on the size of the component to be purified.

Table for ammonium sulphate precipitation (0°C) Use powdered ammonium sulphate and add required weight to get the desired saturation % Saturation (Weight of ammonium sulphate in grams) (I_0 Initial concentration of ammonium sulphate)

I_0	20	25	30	35	40	45	50	55	60	65	70	75	80	85	90	95	100
0	10.7	13.6	16.6	19.7	22.9	26.2	29.5	33.1	36.6	40.4	44.2	48.3	52.3	56.7	61.1	65.9	70.7
5	8.0	10.9	13.9	16.8	20.0	23.2	26.6	30.0	33.6	37.3	41.1	45.0	49.1	53.3	57.8	62.4	67.1
10	5.4	8.2	11.1	14.1	17.1	20.3	23.6	27.0	30.5	34.2	37.9	41.8	45.8	50.0	54.5	58.9	63.6
15	2.6	5.5	8.3	11.3	14.3	17.4	20.7	24.0	27.5	31.0	34.8	38.6	42.6	46.6	51.0	55.5	60.0
20	0.0	2.7	5.6	8.4	11.5	14.5	17.7	21.0	24.4	28.0	31.6	35.4	39.2	43.3	47.6	51.9	56.5
25		0.0	2.7	5.7	8.5	11.7	14.8	18.2	21.4	24.8	28.4	32.1	36.0	40.1	44.2	48.5	52.9
30			0.0	2.8	5.7	8.7	11.9	15.0	18.4	21.7	25.3	28.9	32.8	36.7	40.8	45.1	49.5
35				0.0	2.8	5.8	8.8	12.0	15.3	18.7	22.1	25.8	29.5	33.4	37.4	41.6	45.9
40					0.0	2.9	5.9	9.0	12.2	15.5	19.0	22.5	26.2	30.0	34.0	38.1	42.4
45						0.0	2.9	6.0	9.1	12.5	15.8	19.3	22.9	26.7	30.6	34.7	38.8
50							0.0	3.0	6.1	9.3	12.7	16.1	19.7	23.3	27.2	31.2	35.3
55								0.0	3.0	6.2	9.4	12.9	16.3	20.0	23.8	27.7	31.7
60									0.0	3.1	6.3	9.6	13.1	16.6	20.4	24.2	28.3
65										0.0	3.1	6.4	9.8	13.4	17.0	20.8	24.7
70											0.0	3.2	6.6	10.0	13.6	17.3	21.2
75												0.0	3.2	6.7	10.2	13.9	17.6
80													0.0	3.3	6.8	10.4	14.1
85														0.0	3.4	6.9	10.6
90															0.0	3.4	7.1
95																0.0	3.5
100																	0.0

Chapter 12

General Microbiology

1. Rules and Regulation for Safety in a Microbiology Laboratory

A clean laboratory prevents contamination of experiments, which requires the use of living organism. So it is important to have some precautions, as many of the microbes have pathogencity.

A microbiology student must adhere the following rules.

A. To Reduce the Ever Present Microbial Flora

1. Book and other materials should in specific location, never on bench tops.
2. Door and windows must be closed at working time to prevent contamination by air current.
3. Before and after the experiment clean the bench top by a suitable disinfectant solution.

4. Contaminated instruments like loop, pipette, needle etc. should not be left on the bench tops.
5. Rapid and efficient manipulation of fungal culture to prevent the spreading of reproductive spores.
6. On completion of the practical all cultures and materials should be dispersed properly.

B. To Prevent Accidental Injury and Infection

1. Wash hands with liquid detergents and dry with paper towel upon entering and prior to leaving the laboratory.
2. Wear apron during working to prevent contamination or staining solution.
3. Smoking, drinking, eating prohibited in the laboratory.
4. Cultures should be carried in tray or rack.

C. On Handling Body Fluids of Unknown Origin

Due to the threatening transmission of HIV and Hepatitis B in the unknown test specimen do the following precautions

1. Disposable gloves must be worn during handling of the samples
2. Immediate hand washing on contact with the fluid.
3. Mask, safety goggles should ware as an aerosol might be formed.
4. Test specimen and supplies with these fluids must be placed into disinfectant container prior to autoclaving.

2. Cleaning of Glass Wares

Glass wares are cleaned using chromic acid and detergent. Chromic acid is a chemical used as detergent.

Preparation: A saturated solution of potassium dichromate has to be prepared in water by dissolving 7 gm of potassium dichromate in 35ml of concentrated sulphuric acid with constant stirring. To this add 20 ml of distilled water. Mix well keeping the beaker in ice cold water. Ditrate and use.

Principle: Chromic acid is a strong oxidizing agent to organic material. When glass wares are soaked in chromic acid for 1-2 hrs organic material sticking to it can be removed. Plastic wares are not soaked in chromic acid.

Detergents: Detergents are chemical agents for cleaning purpose (Soap, detergent solutions etc). After removing glass wares from chromic acid they are rinsed with tap water. Then it is thoroughly washed with detergent solution and then with water. Rinse with distilled water 2 to 3 times and dry in a hot air oven.

Principle: Cleaning action of detergents: Detergents come under the class surfactants. These substance effect the energy relationship at interface or intra surface producing reduction in surface interface bond and other particulates of matter adhering to the surface of the glass wares.

Uses: Used for cleaning glass wares, metallic instruments and clothing.

Limitations: Chromic acid can not be used for plastic wares and measuring cylinders.

3. Sterilization

Sterilization by Moist Heating

Principle: Temperature has effect on cellular enzyme system and therefore a marked influence on the rate of chemical reactions and also on life and death of microorganism. Sufficiently low temperature will inactivate and produce a static effect. High temperature destroys cellular enzymes which becomes irreversibly denatured.

Application of heat is common mean of destroying the microorganism. Both dry and moist heat is effective. However moist heat (hydrolyzing and high penetrability) causes coagulation of proteins and kills the cells more rapidly than under dry heat sterilization. The destruction of all forms of life is accompanied in 15 min at 121° C with heat under pressure.

Microbes exhibit differences in their resistance to moist heat. As a general rule bacterial spores requires a temperature above 100° C for destruction where most of the bacterial specimen will be killed at temperature of 50-60° C. Fungal cells require 70-80° C for 10 minutes to denaturize. Common application includes free flowing steam under pressure (autoclaving), free flowing steam at 100° C (tyndillization) and use of lower temperature (pasteurization)

Application of free flowing steam requires exposure of contaminated substances to the temp of 100° C which is achieved by boiling water. Exposure to boiling water for 30 min will result in disinfections. Heat resistant spores may not be killed.

Chapter 13

Media Preparation

Basic requirement of a microbial culture

i) Nutrients

ii) Air/oxygen availability

iii) Moisture

iv) Maintenance of required temperature and pH

v) Preventing contamination

Nutrients: Nutrients are required for the growth and maintenance of the organisms. The general requirements for a microbial growth are

i) *Source of nitrogen :* Nitrogen is derived from the peptones supplied in the medium. Nitrogen is required for synthesizing the cellular components like nucleic acids. Other extracts like yeast extract, beef extract are also good source of nitrogen.

ii) *Source of phosphate and sulphur* : Yeast extract, beef extract and peptone are good source of phosphate and sulphur. These are needed for the synthesis of nucleic acids, phospholipids and some amino acids (methionine, cystine).

iii) *Source of carbon/energy* : Carbohydrates like glucose, sucrose, maltose, and lactose are good source of carbon and energy.

iv) *Inorganic salts and metals* : Metals like sodium, potassium, magnesium, calcium, iron, zinc, cobalt, manganese, and copper are required for microbial growth and supplied as salts to the growth medium.

v) *Water* : Water is an essential component of the protoplasm.

vi) Growth factors like vitamins, amino acids, and purines can be supplied to the medium.

pH: Optimum pH has to be maintained for each type of cultures.

1. Preparation of liquid and solid media

Aim : To prepare liquid and solid media for microbial culture.

Principle : The type of media used in microbiology may vary according to the need of the particular organism under study. Generally media are classified according to their composition.

Complex media : Complex media are used for growing specific microorganism which requires special nutrients for their growth. Generally these media are complex in their composition containing specific nitrogen source, trace elements and complex raw materials like yeast extract, peptone, beef extract etc.

Defined media: Exact chemical composition of the media is defined. It may be composed of entirely organic or inorganic or mixture of both compounds and vitamins. It contains synthetic substances and can be used for specific studies such as metabolic pathways in extraction of metabolites.

Media are again classified as Basal medium, enriched medium selective medium and differential medium.

Basal media: It supports the growth of a wide variety of microorganisms. It can enrich with the addition of some other supplement for a selective growth. Example: Nutrient agar.

Enriched media: This media is enriched with more nutritional requirements especially for the growth of particular organism.

Differential media: It incorporates certain chemical for presumptive identification of a specific organism. It helps in the identification of certain bacteria. Example: TSI agar is specific for gram negative enteric organisms.

Selective media: Selective media can selectively inhibit or support the growth of certain group of bacteria. This is achieved by the use of some chemicals, dye or antibiotic. Example: Mac Conkey agar with bile salt and crystal violet can inhibit the growth of gram positive bacteria while stimulating the growth of gram negative organisms.

Agar is generally used as a solidifying agent. It is a phycocolloid, water soluble poly saccharide obtained from sea weeds or marine algae. Gelidium, garcilaria, agarophytes (a red sea weed), pterocladia etc are used as good source of agar. It is complex carbohydrate composed of mainly D-galactose and 3, 6-anhydro-L-galactose and sulphate as polymers. It is a mixture of two polysaccharides.

Eg. Agarose (neutral polymer) and agaro pectin (acidic polymer). It solidifies just below 45° C and liquefies at 90° C. A medium with agar is called solid medium. A medium without agar is called liquid media or broth.

See APPENDIX A for some common media compositions.

Requirements: Conical flask, cotton plug, autoclave, pH paper, Laminar flow.

LB broth and LB agar.

Procedure: Weigh Peptone-1gm, Yeast extract-0.5 gm and NaCl-1.0 gm

Add all constituents in 100 ml of distilled water. Dissolve and adjust pH to 7.2 to 7.4. Weigh 0.8 gm of agar and add to 50 ml of the medium. Take 50 ml of the medium without agar and 50 ml of the medium with agar in separate 100ml conical flasks and autoclave at 121° C, 15 lb/in^2 for 20 minutes.

Cool and use for inoculation in a laminar flow. For solid medium agar is added after checking the pH. Solid media should be poured to sterile petri plates or test tubes after autoclaving.

2. Isolation of Pure Culture

Aim: To isolate pure colonies of bacteria from soil.

Principle: Microbes in nature are not found growing in isolation but in interaction with the other microbes. However mixed cultures are not very important in study of microbes because it is not possible to determine which organism is responsible for the observed activity. A pure culture is the one containing a single colony of single microbes. This is required to study the concept of growth characteristic, pathogenecity, metabolism and antibiotic reactivity.

The common methods used for the isolation of pure culture are :

1) Streak- plate method

2) Spread - plate method

3) Pour- plate method.

1) *Streak-plate method:* This is done by means of an inoculating loop. A portion of mixed culture is placed on the surface of the sterile medium in petri plate and streaked over the surface. This thin out the bacteria on agar medium so that some individual colony or bacteria may be separated from each other when streaking is properly performed. The bacterial cell will sufficiently far apart in some area of the plate to ensure that the colony developing from all cells will not merge with other growing colony.

2) *Spread - plate method:* A sample of previously diluted bacterial culture is spread over the agar surface using a blend glass rod called spreader.

3) *Pour- plate method:* Multiple dilution of a sample are made and mixed with nutrient broth and poured into Petri plates. These methods are based on the principle that when material containing microbes are cultured, each viable microorganism will develop into a colony. Hence number of colonies appearing in the plate is the index number of the living organism present in the sample.

Procedure

Preparation of soil suspension by serial dilution method: 1 gm of soil is weighed and dissolved in water. Transfer 1ml of soil suspension to test tubes containing 9 ml of water. This process is repeated 3 times. Shake well each time.

Streaking on Petri Plates

1) Streaking was done at aseptic conditions.
2) Inoculation loop is heated to redness, cooled and dipped in soil suspension.
3) Streaked on agar plate without overlapping. Loop was flamed, cooled and streaked on the same plate starting from the end of first streak.

Process is repeated till all area is covered.

Plates were kept inverted to prevent moisture on agar and incubated for 24 to 30 hrs. Number shape and characteristics of the colonies were noted down.

Observation: The following colonies were observed

Rounded colonies:

Irregular shaped colonies:

Yellowish colonies:

Whitish colonies:

Conclusion : Soil suspension has mixed population of bacteria.

Precaution : Aseptic conditions may be maintained throughout.

Streaking should be gentle.

3. Analysis Bacterial Growth Curve

Aim: To study the growth of bacterial cell by plotting a growth curve by taking the optical absorbance of the culture using a colorimeter.

Principle: Bacterial growth is most often related to the number of individual cells in the population. When a bacterium is inoculated into a suitable medium and incubated, the bacterium will undergo binary fission and will result in rapid growth within a period. Here generally the bacterium number will be doubled than the initial number. The time required for doubling of population size is known as doubling time or generation time. This way all population will go through a number of phases that can be represented as growth curve. The growth curve can be divided into the following stages.

1) *Lag phase*: This phase represents a period where there is no increase in cell number. But this phase is referred to as elongation phase of cell and phase of cell adaptations.

2) *Log phase*: This represents the period of time in which cell divides steadily at constant rate and it is also called exponential phase because the number of cells is increasing as an exponential function of time. During log phase the growth rate of bacterium is proportional to the biomass of bacteria i.e. growth rate is proportional to biomass as

   $db/dt = kb$

   where t-time, b-bacterial biomass, k-growth rate constant.

3) *Stationery phase* : In this phase no further increase of growth occurs. In this phase net viable count remain almost unchanged for some time.

4) *Decline phase* : There is decline of number of viable cells.

Requirements : Calorimeter, conical flask with side tube and inoculation loop.

Media composition : Glucose-1.0%, peptone-0.3%, Yeast extract-0.1%, pH-7.2

Procedure

1) Prepare 50 ml of nutrient broth in three conical flasks with side tube.
2) Sterilize the medium.
3) Inoculate with pure bacterial culture (Example- E.coli) in a laminar flow.
4) Take OD at 620nm at an interval of half an hour including 'o' hour.
5) Plot the graph between time on X-axis and OD on Y-axis.

Result: Lag phase, Log phase and stationery phase will be obtained.

4. Study of Effect of Carbon Source on Bacterial Growth

Aim: To study the effect of different carbon source on bacterial growth.

Principle: All forms of life from microorganism to human beings share certain nutritional requirements for growth and normal functioning. All organisms require carbon for synthesizing cell components. All organisms require small amount of CO_2. However some bacteria can utilize CO_2 as their major or sole source of carbon. Such organisms are termed as autotrophs. Others require organic compounds as their carbon source are termed as heterotrophs.

In a medium containing both glucose and lactose, E. coli and some other bacteria use preferably glucose. Lactose is not metabolized until all the glucose is used up. This is known as glucose effect.

Sucrose is the most abundant disaccharides occur through out the plant kingdom.

It can be used by microorganisms which can form glucose and fructose by the action of invertase enzyme.

Starch is a mixture of glucans that plants synthesize and used as a principle food source. It is deposited in the cytoplasm of plant cells as insoluble granules composed of amylopectins. The enzyme α amylase randomly hydrolyze all the α (1-4) glycoside bond of starch except its outermost bond. γ- glucosidase removes one glucose residue at a time from oligosaccharides.

Materials required : Conical flasks, Petri plate, Inoculation loop, Nutrient agar, bacterial sample, spreader.

Media Composition

1) *For glucose as carbon source* : Glucose -1%, Yeast extract-0.1%, Agar 2%, Distilled water-100ml, pH-7.2
2) *For sucrose as carbon source* : Sucrose: 1%, Yeast extract-0.1%, Agar -2%, Distilled water-100ml, pH-7.2
3) *For starch as carbon source* : Starch-1%, Yeast extract-0.1%, Agar 2%, Distilled water-100ml, pH-7.2

Procedure

1) Prepare agar plates with different carbon source.
2) Inoculate the culture on each plate with a spreader.
3) Incubate at 32°C for 24 hrs.
4) Observe for the growth of colonies on each plate.

Observation : Maximum colonies are seen on the plate with glucose as carbon source. Some colonies are seen on agar plate with sucrose and least on plate with starch.

5. Gram Staining (Gram Positive and Negative)

Materials : Colonies of bacteria, Crystal violet, Gram's iodine, 95% ethanol, Safranine, Microscopes with oil immersion.

Procedure

i) Collect the required bacterial suspension. Make a smear on a clean dry slide. Fix the smear to the slide by passing the slide (smear side up) quickly through a flame three times. If the slide is held directly in the flame, it will heat up too rapidly and break. The trick is to gently dry the smear without overheating the slide.

ii) Place the slide on a staining rack. Apply the stains on the fixed smear as follows:

Add 1% crystal violet to the slide and mix well for 30 sec.

Rinse the slide with water.

Add Gram's iodine and mix gently for 60 sec.

Rinse again with water.

Decolorize with 95% ethanol.

Rinse with water.

Counter stain with safranin for 60 sec.

Rinse with water and blot dry.

Examine under oil-immersion objective lens.

Gram-positive bacteria retain crystal violet after washing with 95% ethanol, while gram-negative bacteria lose the purple dye after washing with 95% ethanol. The positive or negative reaction is a measure of the presence or absence of specific polysaccharide components of their cell walls. Safranin is used as a pink counter stain, so that Gram negative cells can be visualized. In practice the distinction is made between purple cells (+) and pink cells (-).

Determine the basic cell shape of the bacteria.

iii) Observe and note down Gram positive and negative stained cells and cell shapes in table.

6. Aseptic Cell Transfers

Materials

Bunsen burner, Wire loop, Petri plate or broth tube, Bacterial culture and Microscope slides.

Procedure

1. Pick up the inoculating loop and hold it pointed down into an open flame until the loop glows red. This process sterilizes the loop of wire and is known as "flaming" the loop. It will result in a sterile loop and will not contaminate your stock culture. If there are liquids already on the loop, the loop should be gradually placed into the flame to dry the loop. If the loop is rushed into the flame, the drop of liquid will splatter and spread bacteria over your work surface.
2. Pick up a broth culture in one hand, while holding the loop in the other. With the last two fingers of the hand holding the loop, remove the cap from the culture and gently flame the top of the test tube (do not overheat). Insert the flamed inoculating loop into the test tube until it is submersed in the broth. The loop should be allowed to cool slightly before immersion. Retract a small quantity of the broth held in the loop and replace the cap on the culture.
3. Open the top of your transfer vessel (tube with water or nutrient agar) and flame the open top of the tube. Insert the loop into the liquid in the tube. Withdraw the inoculating loop slightly from the liquid, blot gently on the inside of the tube and completely remove from the tube. Replace the top on the transfer tube.
4. Immediately flame the inoculating loop.
5. The inoculated tube can be incubated at room temperature or in an incubator for 18 to 24 hrs depending on the culture.

7. Prokaryote Cell Number by Dilution Plating

Materials

Broth culture of E. coli

Tubes of nutrient broth (9.9 ml each)

Nutrient agar plates

Sterile transfer pipettes (1.0 ml)

Procedure

1. Obtain a broth culture of E. coli and carefully mix the contents to ensure equal suspension of the bacteria.
2. Obtain four test tubes each containing 9.9 ml. of nutrient broth. These will be used to produce a serial dilution of the stock culture.
3. Using sterile pipettes remove 0.1 ml of well suspended cells from the stock culture and transfer these aseptically to one of the waiting test tubes of broth. This tube now contains a dilution of 1/100 or 10^{-2}. Gently but thoroughly mix the contents and label this tube as 10^{-2}.
4. Repeat this process, but now aseptically remove 0.1 ml of culture from the 10^{-2} tube and place it into a new broth tube, which now becomes a 1/10,000 or 10^{-4} dilution. Mix the contents and label as 10^{-4}.
5. Repeat this procedure twice more to produce respectively a 10^{-6} and a 10^{-8} dilution. Be sure to mix thoroughly and label each tube.
6. You should now have a serial dilution of the stock culture with tubes at 10^{-2}, 10^{-4}, 10^{-6}, and 10^{-8}. The original stock culture will almost invariably be too high a population for the next step, so use only the four dilutions that have been prepared.

7. Using four separate Petri plates containing 15 ml. of nutrient agar each and four separate sterile pipettes transfer 1.0 ml of each dilution broth suspension on to the surface of a Petri plate. Carefully label the plates and place them in an incubator for 24 hours at 37° C.
8. At the conclusion of incubation, remove each of the four petri plates and count the number of colonies formed on the plates. For proper statistical analysis, the plate containing between 30 and 300 colonies will give the most accurate results.

The colonies can be more easily counted by using a Quebec Colony Counter which allows proper illumination, a grid overlay and by slight magnification of the plate surface. Multiply the number of colonies counted by the dilution factor to obtain the population density of the original broth culture.

Notes A growth curve can be established by repeating this procedure every two hours. Since the number of bacteria can be large, it will be necessary to plate cultures serially diluted. Count the number of colonies for each dilution and average the results.

8. Cell Mass by Measurement of Turbidity

Materials

- Broth (Nutrient agar) culture of E. coli
- Tubes of nutrient agar broth (9.9 ml each)
- Tubes of nutrient agar soy broth (5.0 ml each)
- Nutrient agar plates
- Sterile transfer pipettes
- Spectrophotometer and cuvettes

Procedure

1. Obtain a broth culture of *E. coli*. (Trypticase broth is better than nutrient broth, since it is inherently less light absorbent).

2. Prepare nutrient agar broth dilutions of 10^{-3} and 10^{-6} of an appropriate bacteria culture.
3. Now, for each of the two dilutions, set up a second series of dilutions, this time diluting by 1/2 each time. That is, transfer 5.0 ml of the 10^{-3} culture to 5.0 ml of fresh broth and mix thoroughly. Use 5.0 ml of this and transfer to 5.0 ml of fresh broth to produce a 1/4 dilution. Repeat for 1/8, 1/16, and 1/32 dilutions. Repeat the entire 1/2 dilution series for the 10^{-6} dilution.
4. You should now have twelve tubes, six for each of the 10^{-3} and the 10^{-6} dilutions. These are then diluted 1/2, 1/4, 1/8, 1/16, and 1/32. Use 1.0 ml of each of the twelve dilutions and plate on nutrient agar plates to perform a colony count as in the preceding section.
5. Set a spectrophotometer for 686 nm wavelength and be sure it is turned on and functioning properly. Adjust the absorbance to 0%.
6. Place a cuvette containing nutrient broth into the spectrophotometer and adjust the reading to 0% Absorbance (A). This is the blank for all subsequent measurements.
7. Transfer the remaining contents of the 12 dilutions to spectrophotometer cuvettes and measure the %A (OD value) for each. Compute the absorbance for each sample.
8. Count the number of colonies formed for each sample after 24 hours of incubation. Plot the absorbance of each sample against the plate count for that sample. Once this has been accomplished, you will have a value for computing the cell population number directly by measuring the absorbance of a suspension. This can be more readily measured on a continuous basis. To

do this, grow suspensions directly in cuvette tubes and measure the A at timed intervals. Record a growth curve. The tube can be removed from an incubator at intervals, gently suspended, measured for absorbance and returned immediately to the incubator. Plot cell growth against time to produce a growth curve.

9. Isolation of Bacteriophage from Sewage Sample.

Aim: To isolate bacteriophage from sewage sample.

Principle: Phages have a capsid head and elaborate tail structure. The tail structure allow penetrating the bacterial cell membrane and cell wall to inject their DNA into cell. Phages act as vector carries in transferring DNA from a donor bacterium into recipient bacteria by a process called transduction. Phages are host specific. Phages are very small and detected by holes or plaques. They form lawn of susceptible host bacteria.

Bacteriophages are isolated depending on the formation of plaques formed on lawn of bacterial culture. They present more in soil and in sewages sample. Each plaque is formed by infection of a single bacterium by a single bacteriophage particle. This can then adsorb and infect neighboring bacteria which in turn release another generation of virus particle and after several hours of infection it becomes visible. Each derived from a single plaque are genetically identical to one another.

Materials

Separating funnel, Sewage sample, sterile plates, chloroform, conical flaks, laminar air flow, pipettes.

Media composition
LT agar-1000 ml
Tryptone-10 gm
NaCl-2.5 gm
Agar-2.5%
pH-7.2

Soft agar-100 ml
Tryptone-10 gm
NaCl-2.5 gm
Agar-0.8%

MT broth-100 ml
Maltose-0.25%
$MgSO_4$-10 mM
pH-7.2

Procedure: Collect 20 ml of sewage sample and add 20 ml of chloroform in a separating funnel and mix it. Collect two layers in two different beakers. This is to inactivate the bacteria in chloroform. Take 10 ml of supernatant portion and add to 5ml of MT broth .Allow it to grow for overnight. Prepare agar plates. Take 0.1ml, 0.2 ml, 0.4 ml of overnight grown culture in a tube containing 23ml of soft agar which was maintained at 45° C water bath. Shake well and pour to sterile LT agar plates. Incubate at 37° C for 24 hrs.

Observation: Plaques will be observed.

10. Standard Plate Count

Aim : To determine the purity of water by standard plate count.

Standard plate count is a common procedure for the enumeration of living bacteria. Water consists of some microorganism. In standard plate count the water sample is inoculated on nutrient agar medium by pour - plate method. It will result in a visible colony after incubation. By counting the number of colony forming unit, the concentration of bacteria in the original sample can be determined.

SPC/ML= Dilution factor x Average count.

Limits: Media composition varies from bacteria to bacteria.

Materials

Medium-1000ML

Yeast extract: 2.5 gm

Tryptone: 5 gm

Glucose: 1 gm

Agar: 2%

pH- 7.2

Procedure: Make sterile plates with media of above composition. Pour 1 ml of water samples separately to separate plates. Spread it uniformly. Incubate at 37° C for 24 hrs. Count the number of colonies in each plate.

Result: The number of colonies present

11. IMVIC Test

The difference between E.enterogens and E.coli in citrate utilization can be used as the physiological test and is called IMVIC test. E.coli show positive results for Indole and Methyl red and negative for Voges Proker test and Citrate utilization test. Citrate is only source for carbon and ammonium salt is the only source for nitrogen for Enterobacter aerogenes and E.coli. Organisms which can utilize citrate can grow well in the Simmon's citrate agar medium. Bromothymol blue is added as an indicator in the medium (green in neutral and blue in alkaline medium).

I-Indole test

M-Methyl test

V-Voger proker

C-Citrate utilization test

E.enterogens show positive result for Voger and Citrate utilization test.

Indole test (Tryptophan hydrolysis test)

Certain bacteria such as E.coli have the ability to split the amino acid tryptophan into indole and pyruvic acid by an enzymatic reaction that causes hydrolysis .Indole can be easily detected by Kovac's reagent.

Procedure: Prepare an overnight broth culture of E.coli. Add 10 to 12 drops of Kovac's reagent to the culture in tryptophan broth. A red layer will form at the top of the culture.

Result: Red layer formation is an indication of indole production due to the presence of E.coli.

Methyl red test : A considerable number of gram negative industrial bacteria can be differentiated on the MRVP media and ferment glucose. E.coli, Salmonella, Protease etc. ferment glucose to produce large amount of lactic acid, acetic acid, succinic acid and formic acid etc in addition to CO_2, H_2O and ethanol. The accumulation of acid lowers the pH of the media below 5 or less if methyl red is added to such a culture. Indicator turns to red and indicates that the organism is a mixed acid fermentor. It produces gas due to the production of CO_2 and formic acid.

Procedure : Add few drops of methyl red to the culture. The tube should become red immediately if it is positive.

Result: After adding few drops of methyl red to the culture turns red color immediately. Positive result i.e. Red color due to acid production by *E. coli.*

12. Antibiotic Sensitivity Test on Microbes

Aim : To determine the Minimum Inhibitory Concentration of antibiotics.

Antibiotics are chemotherapeutic agents which inhibit or kill bacterial growth.

Usually antibiotics are metabolic product of one organism and inhibitory to the growth of other organism. Antibiotics have different action mechanism in inhibiting and killing the microbes. This can be divided into five major types

i) By inhibiting cell wall synthesis. Example: penicillin

ii) By inhibiting protein synthesis. Example: streptomycin

iii) By inhibiting nucleic acid synthesis. Example: ciprofloxacin

iv) By reacting with plasma membrane. Example: vancomycin

v) By inhibiting metabolic pathway like folic acid biosynthesis. Example: sulfonamides

Antibiotics have different properties which can be classified according to its effect on microbes i.e. static, cidal, broad spectrum and narrow spectrum. The activity of an antibiotic can be determined by its concentration which is effective in arresting /killing the pathogen. It should have minimum effect on the host cell.

Baur-kirby Test : This is standard method for the determination of antibiotic sensitivity of a microbe. In this the microbe which has to be tested is inoculated on the surface of a sterile Muller-Hinton agar plate. Antibiotic discs were impregnated on the agar plate. On incubating the plates, the inhibition of growth can be noted as the zone of inhibition. The diameter of the zone can be noted and the concentration of the antibiotic to obtain a zone of growth inhibition is known as Minimum Inhibitory Concentration (MIC).The lowest antibiotic concentration which is effective in preventing the growth of the bacterium is known as Minimum Inhibitory Concentration.

Minimum Bactericidal Concentration: MBC. This is the minimum lethal concentration i.e. to kill the microbe.

Materials required: Petri plates, conical flasks, spreader, laminar flow, culture, forceps etc.

Anti biotic disc - penicillin, streptomycin, ampicillin, gentamicin etc.

Media with the following composition can be used for the culture

Glucose-1.0%, peptone-0.5%, Yeast extract-0.1% pH-7.2, Agar-2%

Procedure

Prepare sterile plates with medium.

0.1 ml of any of the overnight culture (eg: *E.coli*, pseudomonas, streptococcus etc.) may be inoculated by spreading on the sterile culture plates. Place the antibiotic discs with different concentration on the agar spread over the plates in the increasing order of concentration. Incubate at 37° C for overnight. Observe the inhibition zone. Note the diameter of the inhibition zone.

Result : MIC obtained for the antibiotic are

1) Penicillin=
2) Streptomycin=

Note : The experiment may be repeated with different antibiotic and different strains.

13. Study of Transformation in Bacteria

Aim: To demonstrate the process of transformation in Bacteria.

Theory: Transfer of naked DNA from one bacterium (donor cell) to another bacterium (recipient cell) through a medium is called transformation. It is a mode of recombination found in several species but not found in all species.

Transformation was first observed by Griffith in 1928 in pathogenic strain of Diplococcus pneumonia. It does not occur in vivo in all species of bacteria and occur in only in those species that posses enzymatic machinery involved in the active uptake of DNA. The recombination processes is normally an in vitro event. Transformation is observed in a wide variety of genera like Bacillus, Haemophilus, Nesseria, and staphylococcus and in the yeast sacchromyces cerviciae.

All the cells in a population are not capable to take naked DNA (naked DNA is not incorporated to the chromosome of the bacteria).Only competent cells posses ability to uptake the DNA which occurs at certain stage in a cell growth cycle, mostly prior to the completion of cell wall synthesis. In this stage a protein called competence factor is released into the medium and it apparently facilitates the entry of DNA. If the competence factor transferred from one culture to the other culture which lacks it cell wall in the treated culture will become competent to receive DNA. Now they can take up DNA fragments.

However not all bacterial cell can become competent and thus not all bacterial cell can be transformed .DNA entry depends on such factor modifications of cell wall and the transformation of specific receptor site on the plasma membrane that can bind DNA .They can bind DNA from the same genus and species or closely related genera and generally reject DNA from distinct related organism.

The process of transformation can be divided into the following steps.

i) Binding of the DNA to the receptor site of the cell surface.

ii) Entry of one strand of DNA across the membrane and cell wall of bacterial cell.

iii) Degradation of other strand by nucleolytic action.

iv) Integration of whole or a part of single strand of donor DNA.

v) Left over DNA is subsequently broken down.

Generally transformation is carried out artificially, where the cells are made to be competent by suspending it in cold $CaCl_2$ and then give a heat shock at 42° C and this is called induced competence.

Transformation efficiency is the number of transformed cells obtained from 1µg of intact plasmid DNA.$CaCl_2$ treatment gives $10^{5 \text{ to } 7}$ transformation of1µg of plasmid.

Principle: Some bacteria have membrane proteins to recognize and facilitate the uptake of short DNA sequence which are not incorporated to *E.coli*.

Actual mechanism of DNA uptake in *E.coli* is unknown but it is supposed that DNA passes through channels formed at adhesion zone where outer and inner membrane fuse to form pores. Since cell surface carries negative charge and DNA has negative charge due to the phosphate group present, cannot approach the cell surface. So in order to neutralize the negative charge, cells are treated with $CaCl_2$ at 0° C which crystallizes all the fluids on the membrane and stabilizing the distribution of negative charges. The Ca 2+ cation in the transformation solution form complexes and shield the negative charges. These plasmid molecules can move through the adhesion zone approaching the cell.

At this point a heat shock at 42° C for 90 sec is given to the cell DNA mixture followed by quick chilling. Heat shock perhaps creates a thermal imbalance on the *E.coli* cell membrane that physically helps to pump DNA through the adhesion zone. After transformation, cells are incubated for 30-60 min. Addition of some LB broth at 37° C and to express antibiotic resistance gene which allow the selection of transformed cells. When transformed E.coli cells are plated on LB agar medium containing antibiotic (ampilicilin) only those which take up DNA i.e. transformed cell grow on antibiotic resistant after transformation. Those are untransformed failed to grow on antibiotic LB agar medium. Ampicillin is a semi synthetic antimicrobial agent used as a selective marker in transformation process.

Requirements

Ampicillin: 100mg/ml distilled water.

Saline solution-1N NaCl

LB medium

Procedure: Prepare 500ml of LB broth in two flasks with one flask containing ampicillin. Pour to plates after sterilization

Prepare overnight E. coli culture (DH 5 α)

Take 100ml of LB broth in a pre cooled sterile centrifuge tube. Centrifuge at 5000 rpm for 10 min at 4° C. Decant supernatant and suspend pellet in sterile 10 ml of 0.1 M $CaCl_2$. Incubate at 0°C for 30 min. Centrifuge at 5000 rpm for 5 min. Re-suspend pellet in 2ml of ice cold 0.1 M $CaCl_2$. Add 10 micrograms of plasmid DNA (pUC18) to 10-20 µl.

Incubate on ice for 30 min. Heat shock for 2min at 42° C in a water bath. Add 1 ml of LB broth to this and incubate for 1 hr at 37° C on a rotary shaker.

After incubation collect 25 and 50 µl of culture and plate it on a petri plate with LB by uniform spreading.

Take inoculums and dilute with sterile PBS 10^6 and 10^8 and inoculate on petri plates with and without ampilicillin. Take 0.1 ml of competent cell without plasmid DNA and inoculate on LB agar as control. Incubate at 24 hrs at 37° C.

Observations: There is no growth on control plate. More growth on the plate inoculated with 50 µl than 25 µl.

Conclusion: Transformed cells can grow successfully on the medium which contain ampicillin and untransformed cell can not grow on the ampicillin medium and normal growth on medium without ampicillin. Plasmid puc18 are resistant to ampicillin and make competent cells resistant to ampicillin.

14. Study of Bacterial Conjugation

Aim: To observe bacterial conjugation

Principle: Conjugation involves the transfer of fertility factor (F plasmid) from donor cell to recipient cell through direct contact and genetic information is passed through. The bacterial cell in which fertility factor is found is known as donor cells or F+ and in which is not present is known as recipient or F- cells. Some plasmids have the ability to combine with the genomic DNA and are called episomes.

Conjugation is a type of genetic recombination placed under para sexual type of reproduction (false sexual reproduction) because sexual reproduction is not possible in the absence of gamete formation. Conjugation was discovered by Lederberg and Tatum in 1946 while they were attempting to cross the different strains of bacteria. The F+ factor injects to the host chromosome and only one of the main sites may be getting integrated to the chromosome and it is known as HFr. In integrated state F- factor mediates the transfer of a chromosome of HFr cell to the recipient F- cell. Usually a portion of the HFr chromosomes are transferred before the cells separate and break the chromosome. It is believed that the transfer of DNA is initiated by an endonucleolytic nick on one strand of specific site known as the origin of transfer on the F- factor. The 5' end of this nick passes through the conjugation tube into the recipient cell. Two strands of a plasmid (F- factor) separate by rolling circle replication process. Only one strand move from one bacterium to the other. Two single strands serve as template for the replication of a complete double strand DNA molecule in both donor and recipient cells. A recipient cell that received DNA as a result of conjugation i.e. called transconjugant. Most of the naturally occurring plasmids are either self transmissible or mobilizable. They have many different system of transfer mechanism. Specific DNA endonucleases known as relaxase in the donor cell which is made up of proteins bound with Ori-T sequence of plasmid. It helps in separating the strands at the Ori-T sequence to initiate the transfer. It makes single break or nick at a specific site called Nick site indicating the transfer process. The relaxase break the phosphodiester bond at the nick site by transferring the bond from a deoxynucleotide to one of the site.

Bacterial cell having F factor have the ability to form sex philes and are called donor cells. When F+ and F- cells are cultured together the donor transfers its F factor to the recipient cells. When strains of E.Coli containing the tetracycline gene is the donor and recipient strains which already contain genes resistant to streptomycin are cultured together will be resulted conjugated cells and it can grow in medium containing both streptomycin and tetracycline.

Materials

Streptomycin- 200 µg/ml,

Normal saline-250 ml,

Sterile media plates.

Amino acids

Threonine, Lucine -20 µg /ml each, Lactose stock - 20 µg/ ml

Prepare minimal media, TBY agar.

Composition of the media is given in APPENDIX A (While preparing minimal media $MgSO_4$ can be separately autoclaved and mixed after).

Procedure

Prepare minimal media broth in 100ml flasks. Sterilize and cool.

Add streptomycin-200 µg/ml.

Add threonine, lucine and lactose at concentration of 20 µg/ml to separate conical flask with 100 ml of minimal medium.

Prepare TBY agar. Take 10 plates. Pour 20 ml of the media to each plate.

Take 0.5 ml of Hfr strains inoculum and add 5 ml of F-inoculum. It is taken in 15 ml of TBY broth and shake. Allow to grow overnight.

Take 0.2 ml of the sample and inoculate in minimal medium of different nutrients at different intervals.

Take 0.05 ml of F- diluted in saline for10^{-4} dilution and pour to plate with TBY agar at different intervals. Similarly different dilutions are inoculated at different intervals of time.

Inoculate Hfr strain to minimal medium with streptomycin which is the control plate. Also keep control for Hfr-strain. Incubate at 37° C for 24 hrs.

Take observation for each plate and each time interval.

Observation

The control plate does not have any growth. The difference in growth is observed for each dilution and time interval.

15. Effect of pH and Temperature on Bacterial Growth

Aim : To study the effect of temperature and pH on microbial growth

Principle

Temperature: Different bacteria have different optimum temperature for growth. This is the temperature at which maximum reproduction occurs. Accordingly the bacteria can be classified as

Psychrophilus-Less than 20° C

Mesophyllus: Moderate temperature-20° C to 40° C

Thermophyllus: Above 40° C

Extermophile: Above 80° C

Procedure: Nutrient agar is prepared and different strains are inoculated and incubated at 4, 10, 20, 30, 35, 40, 50, 60 and 80° C.

The plates were observed for maximum colonies.

Effect of pH: Bacteria have different optimum pH for their growth. Usually bacteria grow in a pH range of 6-9.

Neutrophilus: At pH 7

Acidophilus- pH less than 7

Alkophilus: pH above 7.

Procedure: LB media is prepared at different pH (3, 4, 5, 6, 7, 8, 9 and 10) in petri plates and incubated at 37° C for 24 hrs.

Result: Maximum colonies were obtained at pH

1) For strain no.1 is

Note: Repeat the experiment with different strains.

16. Immobilization of Yeast Cells and Alcohol Production

Aim: To immobilize yeast cells and use the immobilized cells for alcohol production

Principle: Sodium alginate suspension can be used to immobilize yeast cells. Yeast cells when dropped into $CaCl_2$ solution; it forms solid Ca-alginate beads. The beads thus formed will have yeast cells within it. The yeast cells can not pass out through the pours of the beads. But glucose can pass through the pours of the beads and get converted in to alcohol.

Materials

1) Baker's yeast
2) Sodium alginate- 4% (4 gm in 100ml of distilled water, melt it, keep at 50° C till use).

3) $CaCl_2$ solution-0.2 M (2.2198 gm in 100ml of distilled water)-prepare 1000ml.
4) 3% glucose solution-100ml in distilled water.
5) 1 liter beakers, magnetic stirrer, glass rod, test tube, pipettes, Spatula, Distilled water etc.

Procedure: Prepare the sodium alginate suspension. Mix with 4 gm of wet yeast cells. Mix it uniformly. It is then extruded drop by drop to the $CaCl_2$ solution with constant stirring. Keep for 20 min. Beads will be formed and solidified. Filter the solution and collect the beads. Wash thrice with distilled water.

10 good quality beads are collected and take in a test tube. Add 10 ml of 3% glucose solution to the test tube and incubate at room temperature for overnight. Observe next day. The beads will be present at the top of the solution by the evolution of CO_2. This is an indication of alcohol production.

17. Citric Acid Production using Aspergillus Niger

Aim: To produce citric acid using Aspergillus niger.

Principle and Theory: Citric acid is produced by many plants. It is an industrially important product. It is used in many foods, pharmaceutical and cosmetic products. Chemically citric acid is produced using calcium citrate and glycerol in the year 1855.Using biological methods, citric acid was first produced in industrial level in the year 1923.

Citric acid can be produced using many micro organisms such as aspergillus niger. After inoculation citric acid will be produced and will be accumulated in the medium. Using solid state fermentation or by liquid state fermentation, citric acid can be produced.

Materials: Aspergillus niger culture, sterile medium-500ml

Media composition

Ammonium sulphate-0.5gm

Potassium dihydrogen phosphate-0.3gm

Magnesium sulphate-0.25 gm

Copper sulphate-0.03gm

Ferrous sulphate-0.2 gm

Yeast extract-0.1 gm

Glucose-3.0 gm

Distilled water-100ml

pH - 6.0

Procedure: Prepare sterile agar plate with media of above composition. Incubate at room temperature for 75 hrs. After sporulation, inoculate the spores to liquid medium in a sterile atmosphere. Then incubate till mycelium are observed. It will take two to three weeks. Filter and collect the filtrate. Adjust the pH to 7.2 by adding calcium hydroxide. Heat to 80° C. Calcium citrate crystals will be formed. Sulfuric acid can be added to obtain citric acid from calcium citrate.

Result: Calcium citrate precipitate is formed due to the formation of citric acid.

18. Alcohol Production from Yeast and Estimation of Alcohol

Aim: To produce alcohol from yeast cells and estimate the yield of alcohol

Theory: Alcohol can be produced in large scale using fermentation. Different carbon sources can be used for alcohol production.

Materials: Yeast culture, Media with following composition,

Glucose-3%

Yeast extract-0.1 %

pH- 6.5

Prepare 250 ml of the medium in flasks and five media plates. Sterilize.

Procedure: Inoculate yeast cells on agar plates. Incubate at room temperature for overnight. Select a single colony and inoculate to 100ml liquid media with above composition. Incubate for two to three weeks. Collect the medium in centrifuge tubes and centrifuge the medium at 3000rpm for 15 min. Collect the supernatant and use for alcohol estimation.

Estimation of Alcohol using Potassium Dichromate

Prepare a standard curve using absolute alcohol. Pipette 0.1 ml, 0.5 ml, 1.0ml, 2.0 ml, 4.0ml and 6.0ml of alcohol in a series of test tubes. Make up to 7ml with distilled water. Add 3 ml of potassium dichromate reagent to each test tube. Mix well. Incubate for 5 min. Take OD at 600nm.Formation of dark blue color is an indication of the presence of alcohol in the sample. Plot the graph between concentration on X-axis and OD on Y-axis. Calculate the quantity of alcohol using the standard curve.

Preparation of potassium dichromate reagent: Weigh 7.5 gm of potassium dichromate. Dissolve it in concentrated sulphuric acid i.e. add 65 ml of sulphuric acid little by little by keeping the flask in ice cold water. Care should be taken while adding sulphuric acid.

After adding sulfuric acid carefully make up the reagent to 100ml with distilled water.

Note: Microbes are good source of industrially useful fermentation products such as enzymes, antibiotics, alcohol, food additives and biodegrading products.

Table 5.2.8 : List of some fermentation products

Name of the product	Source	Applications
Amylases	Bacillus subtilis	Textile and alcohol industry
Proteases	Aspergillus niger, Bacillus subtilis,	Food, Digestive aid
Trypsin	Pancreas	Medical
Invertase	S. cerevisiae	Food industry
Pectinase	Aspergillus niger	Food industry
Pepsin	Animal stomach	Digestive
Papain	Papaya	Digestive, Medical
Glucose oxidase	Aspergillus niger	Food industry
Lipase	Mold	Digestive
Naringinase	Aspergillus niger	Removal of bitter taste of citrus juice.
Citric acid	Aspergillus niger	Food industry
Fumaric acid	Rhizopus nigricans	Food industry
Lactic acid	Lactobacillus	Food industry
Alcohol	Molasses, Glucose, Saccharomyces cerevisiae	Food industry
Penicillin	Pencillium chrysogenum	Antibiotic
Streptomycin	Streptomyces griseus	Antibiotic
Vitamin B	Streptomyces griseus	Vitamin
Kanamycin	Streptomyces kanamyceticus	Antibiotic
Polymixin -B	Bacillus polymyxa	Antibiotic
Nystatin	Streptomyces noursei	Antifungal
Amphotericin-B	Streptomyces nodosus	Antifungal
Novobiocin	Streptomyces niveus	Antibiotic against antibiotic resistant bacteria

Exercise: Media optimization may be performed for the production of the above industrial products.

Section 6

PLANT BIOTECHNOLOGY

Chapter 14

Plant Tissue Culture

1. Plant Tissue Culture Laboratory

A plant tissue culture laboratory should have cleaning, preparation and sterilization rooms. A clean dust free sterile area has to be maintained for laminar flow, storing of sterile media and glass wares. There should be a culture room with racks for maintaining the cultures. There should be a light source for providing sufficient light intensity for the growth of the plant. The rooms should be cleaned with detergents. Equipments should be properly wiped. Culture racks should be wiped with 70% alcohol before keeping cultures.

2. Cleaning and Sterilization of Glass Wares

The glass wares used in plant tissue culture laboratory should be clean and stable to heat. Good quality borosilicate glass is recommended for plant tissue culture work. Chromic acid can be used for cleaning glass wares. Chromic acid is prepared by dissolving potassium dichromate in concentrated sulphuric acid .The glass wares should soak in chromic acid for 5 hrs. Wash thoroughly in tap water. Then wash with strong detergents. Wash again and rinse in distilled water 3 to 4 times. Dry it in a hot air oven. Washing can be done mechanically using dish washers.

Disposable sterile plastic wares are also available for plant tissue culture work. This will be more costly. In using plastic ware, care must be taken to avoid damage due to chromic acid, solvents such as alcohol, acetone, chloroform etc, and on boiling with dilute sodium carbonate. Cleaned and sterilized glass wares and media should be covered and kept in dust free sterile area.

Laminar air flow : A plant tissue culture should have a laminar air flow for the aseptic handling of sterile media, glass wares and culture work. It has a motor which can blow air which passes through the HEPA (High Efficiency Particulate Air) filter. HEPA filter can remove particles larger than 0.3 μm. This ultra clean air passes through the working space of the laminar flow. Work space and sides of the laminar flow should be wiped with 70% alcohol. UV has to put on half an hour before the work and air flow must be started after putting off UV for 15 min. This removes the ozone layer formed by the UV. After work the flow should be cleaned and kept closed. Laminar air flow should be kept in a clean dust free area.

3. Media Sterilization

The plant tissue culture media contains several nutrients especially sucrose which may support microbial growth. To avoid this microbial growth, media should be properly sterilized. Different sterilization procedures which can be used are

1) *Autoclaving*: The thermally stable media can be sterilized using an autoclave giving the conditions such as 121^0C temperature, 15 psi (lb/in^2) pressure for 20 minutes.

2) *Filter sterilization*: Thermo labile compounds cannot be autoclaved and need filter sterilization. For filter sterilization disposable syringe filters or auto clavable filter assembly with replaceable filter papers are available. Usually filters with 0.2 µm and 0.45 µm pore size and of different diameters are used. Some plant hormones, amino acids, vitamins etc have to be filter sterilized.

Surface sterilization of ex plant : Ex plant are usually collected from field or from plants grown in a garden. It may contain dust and microbes. So the explant must be cleaned and sterilized properly to avoid contamination in culture. Microbes must be killed and there should not be any damage for the explant. For this, explants have to be washed thoroughly in tap water followed by mild detergents and again in tap water and finally distilled water. After this depending on the nature of the explant, a rinse or wipe with 70 % alcohol may be done if necessary. Then the explant can be taken into the laminar flow where it is kept in 0.5 % $HgCl_2$ for 5-8 minutes. Then wash 5 times with sterile water. Drain off the water each time. Keep it on sterile blotting/filter paper kept in a sterile plate with the help of sterile forceps. Ends of the plant material should be removed with sterile scalpel to remove the dead cells if any, due to the diffusion of detergents or sterilization agents used.

4. Media Composition and Media Preparation

Haberlandt in the year 1902 originated the concept of cell culture and he was the first to attempt to cultivate isolated plant cells *in vitro* on an artificial medium. Different part of the plant have different nutritional requirement. Similarly the nutritional requirement varies among plant to plant. In general plant cell requires inorganic salts, plant growth regulators, vitamins, amino acids, carbon source etc for the growth.

Inorganic salts

Macronutrients: Cultured plant cells require continuous supply of inorganic elements such as carbon, hydrogen, oxygen. The elements required in large amounts are termed as macronutrient elements and these include nitrogen, phosphorous, potassium, calcium, magnesium and sulphur. Nitrogen can be added as nitrate or ammonium ions similarly magnesium and sulphur as magnesium sulphate, phosphorous as phosphate, potassium as potassium chloride or nitrate and calcium as chloride or nitrate.

Micronutrients: In addition to macronutrients plant cells also requires traces of certain elements termed as micronutrients. Since these elements require in extremely small quantities it can be prepared as concentrated stock solutions and can be added to the medium as per requirement. Some of the microelements required by the plants are manganese, zinc, cobalt, copper molybdenum, iron, chlorine, iodine etc.

Iron stock : Iron stock is usually prepared separately as chelated form with Na_2EDTA.

Organic Compounds

Vitamins: Vitamins are required in small quantity but are essential for plat growth. It has catalytic function in enzyme systems. Thiamine is an essential vitamin required in plant tissue culture media while nicotinic acid (niacin) and pyridoxine (vitamin B_6) may stimulate growth. Thiamine is added as Thiamine HCl (0.1 to 30 mg/ml).

Ascorbic acid, biotin etc can also be added as per requirement depending on the nature of the explant.

Carbon source: Sucrose or D-glucose can be used as energy and carbon source in plant tissue culture medium. Optimum growth can be obtained for most of the plant using disaccharides as carbon source. In some cases it can be replaced with monosaccharide for obtaining optimum growth.

Plant growth regulators: The common plant growth regulators are auxins and cytokinins.

Auxins: Auxins are a class of compounds that stimulate shoot cells elongation. Examples: Indole 3-Acetic Acid. , á-Napthol acetic acid-NAA, 2, 4-D

Cytokinins: Promotes cell division in plant tissue and regulates growth and development. Cytokinins are mainly amino purine derivatives although there are some exceptions.

Most commonly used cytokinins are BA-Benzyl Adenine, Kinetin, and Zeatin. Kinetin and Benzyl adenine are synthetic compounds.

The ratio and combination between different auxins and cytokinins is the determining factor in cell division. Gibberellins also can be added to cell culture media if required.

Agar is added for solid media in a concentration of 0.8 to 1%.It is also a source of numerous mineral elements in absorbable form.

Considering all these factors several plant tissue culture media are developed with different combination and concentrations and salts and hormones. Exact composition depends on the explant material. Generally MS media (Murashige and Skoog media) can be used as a starting media for plant cell culture. It has relatively high concentration of nitrogen, potassium and ammonium compared to other media. B5 is another common media used and it has been developed by Gamborg's group in the year 1968.

Stock I: Macro nutrients-Major elements (mg/l) - Prepare 100ml-10 x (Add 10 ml to 100ml medium)

Ammonium nitrate-1650 mg

Potassium nitratae-1900mg

Calcium chloride-440mg

Magnesium sulphate-370 mg

Potassium dihydrogen phosphate-170 mg

Stock II: Micro nutrients - Micro elements (mg/l) - Prepare 100ml-10 x (Add 10 ml to 100ml medium)

KI-0.83mg

H_3BO_3-6.2 mg

$MnSO_4$-22.3 mg

$ZnSO_4$-8.6 mg

$NaMoO_4$-0.25 mg

$CuSO_4$-0.025 mg

$CoCl_2$-0.025 mg

Stock III: Fe stock (mg/l) Prepare 100ml-10 x (Add 10 ml to 100ml medium)

Sodium EDTA-37.3 mg

$FeSO_4$-27.8 mg

Stock IV: Vitamin stock (mg/l). Prepare 10ml -100x (Add 1ml to 100 ml medium)

Thiamine HCl-0.1 mg

Niacin (Nicotinic acid)-0.5 mg

Myo-Inositol-100 mg

Pyridoxine-0.5 mg

Glycine-2.0 mg

Auxins (each as separate stocks)

i) 2, 4 D-100 mg/100ml ethanol
ii) NAA-1mg/ml-1N NaOH
iii) IBA-1mg/ml
iv) IAA-I mg/ml

Cytokinins

Kinetin: 1mg/ml (1N HCl. Dissolve in minimum quantity of 1N HCl and make up to 100ml with double distilled water).

ii) BA: 50mg/100ml (1N HCl)

Note : Stocks can be prepared by dissolving in minimum quantity of the solvent and makeup to 100 ml with distilled water. Stocks can be stored in the refrigerating conditions for 2 to 3 months.

Medium no: 1

Medium for callus culture-Prepare 100ml

For 100ml media pipette out the stocks

Stock I-10 ml

Stock II-10 ml

Stock III-10 ml

Stock IV- 1 ml

Sucrose-3%

MS media with 2, 4 D-1.0 mg/lit and kinetin 0.5 mg/lit.

Make up to 100ml with distilled water. Adjust pH -5.6 to 5.8.

Agar : 0.8 %. Melt agar, Pour to culture tubes (Below half of the level of the size of the culture tube). Sterilize at 121^{0}C, 15psi for 20 min.

Ex plant : Carrot.

Table 6.1.4 : Composition of commonly used media in plant tissue culture (Concentration mg/litre of the medium)

Ingredients	MS medium	B5 medium	WH medium	SH medium
Macro elements				
Ammonium nitrate $(NH_4)_2NO_3$	1650			
Ammonium sulphate $(NH_4)_2SO_4$		134		300
Ammonium phosphate(NH4)$_2PO_4$				
Potassium nitrate KNO_3	1900	2500	80	2500
Calcium nitrate $CaNO_3.4H_2O$			288	
Calcium chloride $CaCl_2.2H_2O$	440	150		
Magnesium sulphate$MgSO_4.7\ H_2O$	370	250		
Sodium sulphate $NaSO_4.10\ H_2O$			460	
Potassium phosphate KH_2PO_4	170			
Potassium phosphate $NaH_2PO_4\ H_2O$		150		
Potassium chloride KCl				
Iron stock				
Ferrous sulphate$FeSO_{4.}7\ H_2O$	27.8	27.8	27.8	15
Sodium EDTA Na_2EDTA	33.6	33.6	2.6	20
Micro elements				
Manganese sulphate $MnSO_4\ 4H_2O$	22.3		1.5	10
Manganese sulphate $MnSO_4\ .\ H_2O$		10.0	0.75	1.0
Zinc sulphate $ZnSO_4.2H_2O$	8.6	2.0		5.0
Boric acid H_3BO_3	6.2	3.0		1.0
Potassium iodide KI	0.83	0.75		1.0
Molybdate MoO_4			0.00015	

Contd...

Table 6.1.4 Contd...

Sodium molybdate $NaMoO_4$	0.25	0.25		0.1
Copper sulphate $CuSO_4.5\ H_2O$	0.025	0.025	0.0015	0.2
Cobalt chloride $CoCl_2$	0.025	0.025		0.1
Vitamins				
myo -inositol	100	100		1000
Nicotinic acid	0.5	1.0	0.5	5.0
Pyroxine HCl	0.5	1.0	0.5	5.0
Thiamine HCl	0.1	10.0	0.1	5.0
Glycine	2.0		3.0	
Sucrose	30,000	20,000	20,000	30,000
Plant growth regulators				
Kinetin	0.4-10	0.1		
Indole 3-acetic acid IAA	1-30	0.1-10		0.5
2,4-D				2.0
Agar	0.8-1% (if required)	0.8-1% (if required)	0.8-1% (if required)	
pH	5.7	5.5	5.5	5.8

Medium no: 2

Media for shoot proliferation-(prepare-100ml)

For 100ml media pipette out the stocks

Stock I-10 ml

Stock II-10 ml

Stock III-10 ml

Stock IV- 1 ml

Sucrose-3%

MS media with Kinetin-0.25 mg/l

Make up to 100ml with distilled water .Adjust pH -5.6 to 5.8

Agar : 1.0%. Melt agar, Pour to tubes. Sterilize at 121^0 C, 15psi for 20 min.

Ex plant : Vanilla.

Medium no: 3

Media for Shoot tip culture. (Prepare 100 ml)

For 100 ml media pipette out the stocks

Stock I-10 ml

Stock II-10 ml

Stock III-10 ml

Stock IV- 1 ml

MS media with 3ppmBA and 1ppm kinetin (Add 0.3 ml BA (1mg/ml) and 0.1 ml kinetin (1mg/ml) to 100ml medium)

Sucrose-3%

Make up to 100ml with distilled water .Adjust pH-5.6 to 5.8

Agar : 1.0%. Melt agar, Pour to tubes. Sterilize at 121^0 C, 15psi for 20 min.

Ex plant : Vanilla

Medium no : 4

Medium for Anther culture (prepare 100ml)

For 100ml media pipette out the stocks

Stock I-10 ml

Stock II-10 ml

Stock III-10 ml

Stock IV- 1 ml

MS media with 1ppmBA and 1ppm NAA (Add 0.1 ml BA (1mg/ml) and 0.1 ml NAA (1mg/ml) to 100ml medium)

Sucrose-3%

Make up to 100ml with distilled water. Adjust pH -5.6 to 5.8.

Agar : 1.0%. Melt agar, Pour the medium to tubes. Sterilize at 121^0 C, 15 psi for 20 min.

Ex plant : Datura anther

Medium no. 5

Medium for vanilla rooting

For 100ml media pipette out the stocks

Stock I-10 ml

Stock II-10 ml

Stock III-10 ml

Stock IV- 1 ml

½ MS media with 1ppmBA and 1ppm IBA (Add 0.1 ml BA (1mg/ml) and 0.1 ml IBA (1mg/ml) to 100ml medium)

Sucrose-3%

Make up to 100ml with distilled water .Adjust pH -5.6 to 5.8

Agar : 1.0%. Melt agar, Pour to tubes. Sterilize at 121° C, 15 psi for 20 min.

Medium no. 6

Medium for carrot callus regeneration

For 100ml media pipette out the stocks

Stock I-10 ml

Stock II-10 ml

Stock III-10 ml

Stock IV- 1 ml

½ MS media with 1ppmBA and 1ppm 1 kinetin (Add 0.1 ml BA (1mg/ml) and 0.1 ml Kinetin (1mg/ml) to 100ml medium)

Sucrose-3%

Make up to 100ml with distilled water .Adjust pH -5.6 to 5.8

Agar : 1.0%. Melt agar, Pour to tubes. Sterilize at 121^0 C, 15 psi for 20 min.

Note : Media composition has to be decided depending on the ex plant. Poly Vinyl Pyrrolidine (PVP) has to be added for the explants with phenolic exudates.

5. Callus Culture in Carrot

1) Aim: To Initiate Callus Growth in Carrot

Principle: Plant cells are totipotent. i. e. any plant cell can divide and develop into the whole plant. Callus is a mass of cells originated from proliferating cells. This can be induced *in vitro* by tissue culture techniques. It has three developmental stages

1) Initiation
2) Division and
3) Differentiation

These stages depend on culture conditions, age and nature of the ex plant etc.2, 4-D is a common synthetic auxin used for callus proliferation. Callus is a mass of cells loosely arranged parenchyma cells arising from the proliferating cells of parent tissue. Plant materials generally used for callus induction are vascular cambia, storage parenchyma, pericycles of root, leaf mesophyll, and provascular tissue.

Somatic embryogenesis : The formation of embryoids from somatic/vegetative cells is known as somatic embryogenesis. Generally somatic embryoids can be obtained from vegetative cells of mature plants, hypocotyls, cotyledons, and reproductive tissues other than zygote. Somatic embryogenesis can be through callus or without callus intermediate.

Materials required : Carrot, $HgCl_2$-0.1%, sterile water, Petri plates, forceps, scalpel, blade, cotton, Medium no 1, (5tubes), detergent, laminar air flow, alcohol-70% etc.

Procedure : Wash the carrot thoroughly in tap water, then with detergent. Cut the carrot into small pieces by removing the sides. Care should be taken to collect the cambial ring and both inside and outside of the cambial ring together to get the actively dividing cells. Cut into small cubes (1cm) of carrot and again wash with distilled water and then with 1%detergent. Again wash it with distilled water and bring into laminar flow. Add 0.1% $HgCl_2$ and keep it for 6min.Wash with sterile distilled water for 3 times. Take out single piece in a sterile filter paper kept in a sterile plate. Cut the sides again to remove dead cells and directly place on the agar surface .Gently press it. Incubate at 24^0C at 0 lux intensity of light for one month in a sterile atmosphere.

Observation : Take observation every day.

Results : Callus initiation started in third week. Sub culture the grown callus to medium no. 7 for shoot initiation or to the medium with same composition for further callus production.

For organogenesis from callus, different auxin: cytokinin ratio can be tried depending on the nature of the explant. A relative high auxin: cytokinin ratio induced root formation, whereas low ratio between auxin and cytokinins favoured shoot formation.

Note: For callus production from leaves, collect young fresh leaves. Clean and wash thoroughly in tap water. Wipe with 70% alcohol and wash with distilled water. Treat with 0.5% $HgCl_2$ for 5 to 10 min. and wash with sterile water for 4-5 times. After surface sterilization, the leaves are brought into the laminar flow and cut into small pieces with the help of a sterile scalpel. These leaf bits are then inoculated into the callus culture medium and incubated in dark for callus production.

Callus can be sub cultured to suspension (liquid) cultures for the isolation of secondary metabolites. Depending on the explant sub culturing can be done to suspension culture for better yield.

Exercise: Try different plants for callus production.

6. Culture of Shoot Tip (Vanilla)

Aim : In vitro culturing of shoot tip for multiplication.

Principle : Shoot tips and nodes have auxiliary bud primordial which can develop into shoots and can multiply in a suitable medium with growth regulators. Cytokinins are added as kinetin or BA, IPA or Zeatin. Auxins such as IAA, NAA or IBA are also required in the meristem culture medium.

Materials required: Vanilla shoots,$HgCl_2$-0.1%,Sterile water, petri plates, forceps, scalpel, blade, cotton, Medium no 2-5 tubes, detergent, laminar air flow,alcohol-70% etc.

Procedure: Collect shoot tips of vanilla, wash thoroughly with tap water and then with 5% detergent. Collect shoot apex. Wash again in a conical flask. Wash with distilled water for 3-4 times. Bring to the laminar flow. Keep in 0.1 % $HgCl_2$ for 6-10min.Wash with sterile water for 3-4 times with stirring. Take one shoot tip on the petri plate. Dry it. Cut the bottom and sides. Insert on the agar. Incubate at 24^{0}C at 1000 lux and 16 hrs photo period for one month in a sterile atmosphere.

Observation: Take observation every day for 1-2 months.

Results: Sub cultured for rooting in medium no.6

Note: Depending on the ex plant changes may be made in the media composition.

PVP may be added to prevent the accumulation of phenolic exudates in the culture.

For example when pepper is used as ex plant phenolic exudates will ooze out to the medium. This has adverse effect on the growth of the plant. In order to prevent this PVP at a concentration of 1mg/litre can be added to the culture medium.

7. Anther Culture of Datura

Aim: To produce haploid plantlets by anther culture.

Principle: Anthers are haploid cells and can regenerate into haploid plants if suitable culture conditions are provided. The main advantages of haploid plants are that they can produce seedless fruits. Example-grapes, watermelon etc.

Materials required : Datura buds, $HgCl_2$-0.1%, Sterile water, Petri plates, forceps, scalpel, blade, cotton, Medium no 2, (5 tubes), detergent, laminar air flow,alcohol-70% etc.

Procedure : Collect mature buds. Wash and wipe with 70% alcohol. Open the bud carefully inside the laminar flow. Collect the anther. Place it on the agar surface. Gently press it.

Incubate at 24^0C at 0 lux and 16 hrs photo period for one month in a sterile atmosphere.

Observation : Take observation every day for 1-2 months.

Results : Callus can be observed after one month.

Exercise : Practice and use different ex plant material for inoculation. Take observations and optimize the media composition.

8. Subculture of Carrot Callus

Cultures should be sub cultured because the nutrients may have exhausted, depleted, dried up. Timely sub culturing must be done for the proper maintenance of the cultures.

Materials required : Carrot callus cultures, MS medium with 3ppm BA and 1 ppm IAA, Laminar air flow etc.

Procedure: Take out the callus from the tube/jar to a sterile plate kept in a laminar air flow. Large callus can be cut into small pieces by a sterile scalpel .Place it in the sub culturing medium and incubate at 25 ^{0}C and 16 hrs photoperiod 1000 lux for shoot regeneration.

Note: Sub culturing of vanilla shoot tip cultures: The developed shoot multiples can cut into single ones and can be sub cultured for rooting in a medium with 3ppm BA and 1ppm IBA.

9. Hardening of Plantlets

Tissue culture plants are grown in vitro in suitable conditions for the growth. So in order to transfer the plantlets to the field, a conditioning called hardening is required. This is done for the acclimatization of the plantlets to the environmental stress. For this the plantlets are kept in a green house/hardening unit where artificial humidity, temperature etc are provided. The plantlets are taken out from the tissue culture media and kept in sterile soil in small pots. MS basal media can be sprayed daily. It is kept for one week to 10 days in the hardening unit and survived plantlets can be transferred to the sterile soil and can be kept in external atmosphere in a net house till it grow for some time and transfer to field.

10. Isolation of Protoplast from Plant Leaves

Aim: To isolate protoplast from leaf cells.

Principle: Cell wall can be removed by enzymatic method and plant cell without) cell wall is the protoplast. Macerozyme R-10 (Pectinase), cellulase (cellulase Onozuka R-10), hemicellulase etc can be used to remove the cell wall and release protoplast.

Materials: Leaves of pepper (colubrinum), Datura.

Salt solution: Cell Protoplast Washing (CPW) solution: KH_2PO_4-27.2 mg/l, KNO_3-101.0mg/l, $CaCl_2$. H_2O - 1480mg/l, $MgSO_4.7H_2O$-246mg/l,KI-0.16mg/l,$CuSO_4$. $5H_2O$-0.25mg/l,pH-5.8.

Protoplast culture medium: MS basal +sucrose 3% +Mannitol-9% +1 mg/l NAA +0.2 mg/l 2, 4-D+0.5 mg/L BA, pH-5.8.

Enzyme solution: Cellulase-0.5%, Marcerozyme-0.25%, MES Buffer-2Mm, Mannitol-9.0 gm, CPW as required. pH 5.4 to 5.8.

Mannitol-13 %

Sucrose 21%

Procedure: Collect 5-6 weeks old well expanded leaves.

Remove lower epidermis by peeling off.

Cut into small thin pieces by a scalpel blade.

Transfer the leaf pieces into the CPW13M (CPW with 13%Mannitol) solution in a petri dish (10ml) with peeled surface exposed to the medium for 30 min.

Pipette out CPW 13M solution beneath the lower surface of the leaflets and add enzyme solution.

Incubate for 5hrs in dark at 25^0 C. (3 hrs for Datura, 6 hrs for colobrinum).

Sieve the protoplast solution and centrifuge at 800 rpm for 8 min. Remove the supernatant and add 10 ml of CPW with 21 %sucrose.

Centrifuge at 2000rpm for 5 min. Protoplast will form a band on the top. Collect the band.

Add equal volume of CPW with 13 % mannitol. Centrifuge at 2000 rpm for 5 min.

Collect pellet. Wash with CPW

Transfer to 5 ml of culture medium. Incubate at 25^{0}C in dark condition in small petri dish.

Take 1 ml of solution and count the protoplast using a haemocytometer under microscope.

Observation : Protoplast is observed as green shining globules.

Exercise: Use different ex plant materials and standardize the procedure for each plant.

Chapter 15

Plant Molecular Biology

1. Isolation of Plant DNA

Aim: To isolate DNA from plant leaves.

Principle: In order to isolate DNA from plant leaves, the cell wall has to be removed and other cell components have to be separated and removed. CTAB (cetyl trimethyl ammonium bromide) is a common cationic detergent used to precipitate nucleic acid from solution of low ionic strength. But at high ionic strength CTAB forms complexes with proteins polysaccharides but will not precipitate nucleic acid.

The pH of the buffer can be maintained in between 7 to 8 to maintain the integrity of the DNA. Isopropanol is added to precipitate the DNA. It is faster than the ethanol in precipitating the DNA. EDTA is added to chelate with metallic ions which prevents the activity of nucleosomes and in turn the denaturing of the DNA. Iso amyl alcohol is added to prevent the foam formation during the DNA isolation. PVP and β-mercapto ethanol are added to prevent the oxidation of the phenolic compounds.

Materials : TE Buffer- Tris buffer10 mM pH-8.0, Tris pH-8

EDTA-2.0Mm

Chloroform : Iso amyl alcohol-24:1

Iso propanol (cold)

CTAB extraction buffer : Tris-0.1 M (pH-7.5)

Na Cl-1M

EDTA-0.01M

CTAB (n- cetyl tri ammonium bromide): 1%

PVP-2%

Mercaptoethanol-6%

Ethanol-70%

Procedure : Collect 2gm of young fresh pepper leaves, wash, wipe, and grind it with liquid nitrogen. Take 500 mg from the grinded powder and mix with 5 ml of extraction buffer.

Add equal volumes of chloroform and iso amyl alcohol

Centrifuge at 1000rpm for 5 min.

Collect supernatant. Add equal volume of isopropanol (Cold in eppendorf tubes). DNA will be precipitated.

Centrifuge at 2000rpm for 5min.

Supernatant discarded and pellet dissolve in TE.

Precipitate and wash with ethanol.

Store in TE.

Check OD ratio 260/280 (1.8 for pure DNA isolation, less than 1.8 is due to protein contamination).

The DNA can be stored in TE buffer and can be used for electrophoresis, Restriction digestion and RAPD analysis.

Exercise: Isolate DNA from different plants. Add antioxidants in suitable concentration if required.

2. RAPD Analysis of Genomic DNA

Aim : To perform the **RAPD** of the isolated plant genomic DNA.

Principle : Random Amplified Polymorphic DNA can be obtained using Polymerized Chain Reaction (PCR). Genomic DNA of closely related species have few DNA difference along the length of the DNA. This difference may be any where in the genome. Due to this difference in the genomic DNA of two species, when the DNA is amplified by PCR using random primers products of different banding will be formed on the gel. Most of the bands will be similar, but there will be some different bands which indicate the difference in species for the same primers. Usually decamers are used as primers and the results will be the same on repetition.

Materials

Samples of genomic DNA

PCR buffer-10 X

dNTPs,

Taq polymers

Primers

Thermocycler

Procedure : Five sample of DNA from different sources with different concentrations such as 0.5 μg/ml,1.0 μg/ml, 1.5 μg / ml are collected. Dilute the DNA sample to get concentration of DNA as10ng/ml.

25 µl of reaction mix are prepared with the samples as follows

Template DNA-2.5 µl (25ng) +Buffer (10X) - 2.5 µl + dNTPs - 1.5 µl+ primers+ sterile water-18 µl+ Enzyme (Taq polymerase)-3U-0.5 µl. Mix well by spinning in a PCR tube and keep in PCR machine. For running the PCR the cycles are as follows:

I

1 CYCLE-94°C for 5 min (Denaturation)

II

10 cycles each cycles with

i) 94^{0}C for 45 sec

ii) 35^{0}Cfor 1min

iii) 72^{0}Cfor 2 min

III

45 cycles of 3 steps

i) 94^{0}C for 40 sec

ii) 38^{0}Cfor 1min

iii) 72^{0}Cfor 2 min

IV

72^0 C for 10 min.

After completion of PCR, samples are allowed to cool and load on agarose gel to run. Markers are also loaded in one well. Observe the bands. Analyze using suitable software.

Results: Similar bands are observed as well as different bands.

Note: Bioinformatics tools can be used for genome analysis and search. Primer designing for any genes is based on the database search and is available in the PRIMER3 site (http://www.frodo.wi.mit.edu).

3. Agro Bacterium Based Transformation in Plants

Aim : To obtain transformed plants using agro bacterium tumefaciens.

Principle : Agro bacterium is a soil born bacteria which can infect plant cells causing the formation of crown gall diseases. It can infect the plant cells by DNA transfer and integrating into the plant genome. The DNA segment which is transformed is called T-DNA. The T-DNA flanked by 26-bp direct repeats and the DNA lies between these repeats is transferred and integrated in the plant genome.

Callus cultures, protoplast cultures, suspension culture etc can be used as ex plant for the transformation. For the transformation of the desired gene, the same may be cloned to agro bacterium and can be used for the infection of the ex plant.

Selection of transformants : Usually the agro bacterium with selectable marker is co cultured with explant in the medium with suitable antibiotic in toxic level, where the agro bacterium will be killed and only transformed plant will survive. This can be sub cultured for regeneration in a suitable medium.

Materials required : LB medium, sterile plates, MS basal medium, kanamycin, sterile water, forceps, and laminar air flow.

Procedure : Prepare LB broth and LB agar .Inoculate Agro bacterium and prepare suspension cultures by incubating for 24 hrs. Centrifuge at 5000 rpm for 10 min. Collect the pellet. Wash with 10 ml MS basal medium. Resuspend in 20 ml of MS basal medium. Collect the explant (Pepper leaves). Do surface sterilization and bring inside the laminar flow. Cut the surface sterilized ex plant into small pieces. Immerse the ex plant in the agro bacterium suspension in MS medium in a sterile plate. Incubate for

20-30 min. Air dry it by keeping it on sterile filter paper. Transfer to MS basal medium and keep in dark for 2 to 3 days. This is for the agro bacterium infection of the explant. The infected/transformed plant can now place in MS basal medium with suitable antibiotic (Cefotaxime and Kanamycin-50 μg/ml and 200 μg/ml). Repeated subculture will help in the complete elimination of the agro bacterium growing in the medium. Sub culture the above for shoot regeneration in MS medium with suitable selection antibiotic.

Note: The selection antibiotic will vary for different plant. Each plant has to be screened for the antibiotic sensitivity and concentration.

4. Assay for GUS Operon

Aim : To perform assay for GUS operon in transformed plant.

Principle: GUS operon in E.coli consists of the genes

1) GUS A: Codes for β-glucoronidase
2) GUS B: Codes for glucoronide permease
3) GUS C: Function not clear

X-Glu is the substrate used for the GUS gene. X- glu is chemically 5-Bromo4-chloro3-endyl β-D glucoronide. It produces a dark blue cleavage product.

Materials required: X-Glu, transformed plant.

X Glu stock-1mg/1ml of DMSO

Mix 0.5 ml of X –Glu stock, 1.4 ml phosphate buffer, 0.1 ml of tritron/ ethanol. Store at 4^0C.

Procedure: Collect transformed and untransformed plant (control) material. Incubate the tissue in the stain at 37 ^{0}C for 24 hrs. Decolorize the tissue using 90% ethanol. Observe under microscope.

Observation: Transformed plant has blue color indicating the successful transformation of GUS operon.

5. Confirmation of Transformation using Polymerized Chain Reaction (PCR)

Aim : To confirm the transformation using PCR for npt II gene amplification.

Principle : PCR can be used to amplify npt II gene

Procedure: PCR is done in the following conditions using nptII primers

Primer 1: 5′ GTCATCTCACCTTGCTCCTGCC 3′and

Primer 2: 5′GTCGCTTGGTCGGTCATTTCG 3′.

PCR conditions can be given as 95^{0}C for 45 sec, 57^{0}C for 40sec, and 72^{0}C for 2min. Repeat in 40 cycles.

Run PCR product on the agarose gel and confirm the amplification.

Note: NCBI home page (www.ncbi.homepage) can be used for genome search. Information about a gene of interest will be obtained on searching. Using Primer 3 site, primers for a particular gene sequence will be obtained which can be used for PCR amplification studies.

6. Primer Designing

Aim: To use a database for designing specific primers for PCR amplification

Steps: Go to NCBI home page on web (http://www.ncbi.nlm.nih.gov). Select the option gene for a specific enzyme synthesis (For example: give option as chalcone synthase gene in plants). Then page on all sequence details with accession number will appear. Select the sequence from it. Use this sequence for primer designing. PRIMER3 can be used for primer designing. Do PCR amplification using these primers. Note that the Tm is not below 55°C.

7. Isolation, Purification and Identification of Plant Secondary Metabolites

Aim: To prepare a plant extract and analysis of the compounds.

Extract preparation: Collect the plant leaves, bark or any other suitable plant material, wash in tap water, rinse with distilled water. Air dry the material. Powder it. Take 5 to 10 gm of the powder. Cover in grade no1 filter paper. Keep in a soxhlet apparatus. Add the solvent to a level just above the plant material. Switch on the water supply for circulating the water for cooling. Heating mantle should be adjusted to the required temperature of the solvent. Do the refluxing till the solvent is colorless. A sequential extraction can be done with the same plant material in an increasing order of the polarity of the solvent.Collect the extract, evaporate the solvent using a evaporator and weigh the mass obtained. Dissolve in a suitable solvent and store in cold conditions. It can be used as the crude extract for various bioassays and analysis.

Note: The extraction can be done in room temperature also. For this, a known a quantity of the dried plant powder is kept in the solvent with constant stirring for 10 to 24 hrs. Collect the supernatant. Filter through the blotting paper. Centrifuge at 5000 rpm for 15 min and collect the supernatant. This can be evaporated to dryness. Collect the dried material and weigh. Calculate the %yield.

The extracted matter can then dissolve in minimum quantity of solvent or any other suitable solvent required for other bioassays.

The extract can subject to Thin Layer Chromatography (TLC) for further separation and analysis of components.

List of solvent system and spray reagents for the detection plant secondary metabolite is given in the table below.

Table 6.2 : Solvent system and spray reagents for different components of plant extracts

Extract	Component	Solvent system	Ratio of solvent system	Spray reagent
Methanol	Terpenoids	Methylene dichloride: chloroform: ethyl acetate: n-propanol	47.5:45:2:4:5	Vanillin sulphuric acid reagent,105^{0}C,5 min
	Triterpenoids	n-butanol:2M NH_4OH	1:1	Antimony tri chloride- purplish color will be developed.
	Sterols	Chloroform: methanol	3:4	p-ani saldehyde-blue to faint pink spot
	Flavanoids	Toluene: chloroform: acetone	40:25:35	Fc reagent- yellow to orange spot
	Quinones	Ethyl acetate: methanol: water	100:16.5:13.5	Ethanolic KOH. Orange colored spot.
	Alkaloids	Toluene :ethyl acetate: diethyl amine	70:20:10	Dragandorff reagent-orange spots
	glycosides	Chloroform: methanol: water	70:30:5	0.2%orcinol in 2N H_2SO_4
	Saponins	Chloroform: methanol: water	64:50:10	1%vanillin in sulphuric acid
Acetone	Carotenoids	Petroleum ether: Acetone	7:3	Visible light

Note: For reagent preparations see Appendix C

8. Evaluation of Antioxidant Activity of Plant Extract

Phopho molybdenum method: The assay is based on the reduction of Mo (6) to Mo (5) by the extract which is indicated by the presence of green colored complex. Take the methanol extract at a concentration of 1.0 mg / ml. Add equal volume of phosphomolybdate reagent to the extract, mix well and take the absorbance at 695 nm. An increase in OD at 695 nm is the measure of the antioxidant activity.

Reducing power method: Antioxidant compound can form a colored compound when react with potassium ferricyanide, trichloro acetic acid and ferric chloride. An increase in OD at 700 nm is an indication of antioxidant compound.

Take 1ml of the extract with concentration 100 to 1000 μg/ml in distilled water.

Mix with 2.5 ml of phosphate buffer (0.2 M, pH 6.6)

Add 2.5 ml of potassium ferricyanide (1%).Mix well. Incubate at 50^{0}C for 30 min.

Add 2.5 ml of 10%acetic acid in distilled water. Mix well.

Centrifuge at 3000 rpm for 10 minutes.

Collect 2.5 ml of supernatant.

Add 2.5 ml of distilled water

Add 0.5 ml of 0.1% $FeCl_3$ in distilled water. Mix well. Immediately take OD at 700nm for 20 min in an interval of 1 min. An increase in OD is an indication of antioxidant activity of the compound.

DPPH method (2', 2' diphenyl -1-2-di picryl hyrazyl): Anti oxidants scavenge the stable DPPH radical .It can then reduce to hydrazine when it reacts with a hydrogen donor. Dissolve DPPH in methanol to get a 0.1 % solution and add 5 micro liters of the sample to it. DPPH has to be prepared fresh. Sample is prepared with the concentration of 0.1 mg/ml in methanol. Measure OD immediately. Decrease in OD at 550 nm is an indication of the antioxidant activity. Quercetin can be used as standard.

9. Estimation of Total Phenolic Content in a Plant Extract

Aim: To estimate the total phenolic content of the given plant extract.

Phenolic content in a plant extract can be determined by using Folin- Ciocalteu (FC reagent) reagent.

Materials: Plant extract at concentration of 1mg/ml, Spectrophotometer, methanol, cuvettes, distilled water etc.

Procedure: Take 1ml (0.1mg/ml) of the extract and mix with 5ml of distilled water. Add 250 µl of FC reagent (1N) to this. Keep for 3 min at 25 ^{0}C. Add 1ml of saturated sodium carbonate solution. Mix well and incubate for 1 hr at room temperature. Read OD at 760 nm. Use methanol as blank. Plot a standard calibration curve using gallic acid as standard. Calculate the quantity of phenolic components from the standard curve.

Results: Express the results as gram of Gallic Acid Equivalents per 100g of lyophilized extract.

10. Estimation of Total Flavanoid Content

Aim: To obtain the total flavanoid content of the plant extract

Total flavanoids of the plant extract can be estimated using aluminum trichloride.

Materials: Plant extract-2mg/ml, $AlCl_3$ in methanol-2%, spectrophotometer.

Procedure: Mix 2ml of plant extract with 2 ml of the 2% aluminium chloride in methanol. Mix well and take OD at 415nm after 10 min. Take 2ml of plant extract and 2 ml of methanol as blank. Quercetin can be used as standard and plot the standard curve.

Result: The mean of three reading can be expressed as quercetin equivalents per 100g of the extract.

Section 7

ANIMAL BIOTECHNOLOGY

Chapter 16

General Requirements and Experiments in Animal Biotechnology

1. Animal Biotechnology Laboratory

General requirements: An Animal Biotechnology should have storage, preparation and sterilization rooms. Culture room should be free of dust and contaminations. All sterile items are to be stored in sterile area.

Equipments : Laminar air flow,CO_2 incubators, CO_2 cylinders, Liquid nitrogen , Balances of different model, pH meter, osmometer, microscopes, haemocytometer, cell counter, magnetic stirrers, vacuum pump, autoclave, sterile glass wares, disposable plastic wares like petri dishes, culture flasks, syringes, filters, pipettes (all different size),micromanipulators etc.

Common Media in Animal Biotechnology : Most of the animal tissue culture media are readily available and can be reconstituted according to the culture requirements.

Common media are

1) Balanced Salt Solution-BSS
2) PBS-Phosphate Buffered Saline
3) Earl's BSS
4) Dulbecco's PBS
5) Hank's PBS
6) Minimal Essential Medium-MEM
7) Eagle's MEM

A complete media contains all the essential requirements for the growth of the cell.

The major constituents of most of the media are

1) *Amino acids*: All essential amino acids, non essential amino acids as per requirement to obtain normal growth.
2) *Vitamins*: Water soluble vitamins. Composition varies from media to media. Some of the vitamins are derived from the serum added.
3) *Salts*: Sodium, potassium, calcium, chlorine, sulphate, bicarbonate.
4) *Glucose*: Is the source of energy.
5) *Hormones and growth factors:* They are added to according to the type of culture to the medium. Example: Insulin which promotes uptake of glucose.
6) *Antibiotics*: Generally added to avoid contamination and degeneration of culture cells.
7) *Serum supplements*: Are good source of the growth factors for proliferation, minerals, lipids and hormones. Example: Fetal calf serum, Bovine serum, Calf serum etc.

Media are to be selected according to the type of culture.

For example like

CT3 cells-MEM, DMEM

Chick embryo fibroblasts-Eagle's MEM

Fibroblast -Eagle's MEM

Human tumors-RPMI 1640

Mouse embryo-Eagle's MEM

Neurons-DMEM

Skeletal muscle-DMEM, HAM'S F-12

Cattle embryos-TC-199 supplemented with 10% FCS and co - culturing with oviductal epithelial cells.

2. Growing Cells with Serum-Basal Medium Eagle (BME), MEM

High growth density-DMEM

Mice, Hamster Ovary cells, oocytes-HAM'S F-12, Nutrient mixture-low density

Rat liver epithelial cells-Williams medium E

Human leukemic cells, hybridomas-RPM1 1640

Mouse leukemic cells-RPM1 1630

Neuro basal medium-CNS neurons

Chick embryo fibroblast-Medium 199

Note: Sometimes addition of some components like growth factors, hormones, vitamins, amino acids etc can improve the growth of the cells. So media have to be standardized for any particular cell lines. Media optimization can be performed by dose dependent curve on each component in the media. Select the media with optimal range and retest each component.

3. Media Preparation and Sterilization

Glass wares : All glass wares used in animal tissue culture should be of good quality to avoid leaching of components to the medium, which can be toxic to the cells.

Glass wares should be cleaned using proper detergents. After washing it in soap solution wash thoroughly in tap water, rinse in distilled water 4 to 5 times. Dry it in hot air oven.

Media components have to be sterilized according to the nature of components, like

Amino acids-Filter sterilization

Antibiotics-Filter sterilization

BSA- Filter sterilization

Glucose-Autoclave

HEPES- Autoclave

Growth factors- Filter sterilization

$NaHCO_3$- Filter sterilization

Na OH- Filter sterilization

Phenol red-Autoclave

Glutamine- Filter sterilization

Vitamins- Filter sterilization

Trypsin- Filter sterilization

Serum- Filter sterilization

Pyruvate- Filter sterilization

Preparation of Basal Salt Solution (BSS)

BSS powder is readily available. It can be dissolved in the required quantity of de ionized sterile water. Mix well using a magnetic stirrer. Check pH. Take in suitable container. Cap it. Sterilize at 121^{0}C for 15 to 20 min at 15 lbs (lb/in^2) pressure. Store the medium at room temperature. Add the thermo labile components such as glucose, $NaHCO_3$, correct pH to 7.3. Filter sterilize and use. For animal cell culture media osmolarity has to be checked. It should be 290-300 mOsm.

Note : Commercially available media may be reconstituted; filter sterilized and can be used.

4. Contaminations in Animal Tissue Culture

General Contaminations in Animal Cell Culture

Sterile media, glass wares, laminar air flow etc has to be maintained for the prevention of the contaminations. Appropriate location for culture works, routine cleaning and maintenance, frequent disinfection of all working surfaces etc are required to maintain a contamination free atmosphere. The external surfaces of all materials going into and out of the laminar flow hood should be disinfected.

Microbial Contaminations

Microbial contaminations are common in animal cell culture. A contaminated culture may be immediately sterilized and discarded to avoid spreading of contaminations. A suitable antibiotic can be added to the medium to kill the bacteria if it is in the initial stage. The concentration of the antibiotic should be minimum just for destroying the microbes. It should not be toxic to the culture cells.

Contamination by Mycoplasma

Contamination by mycoplasma can not be detected by naked eye. Over growth may be detected by turbidity in the medium and also by pH change. It can be detected accurately only by Hoechest staining and viewing under microscope. PCR and ELISA techniques etc also can be employed in detecting mycoplasma. The mycoplasma can also be detected by culturing in mycoplasma growth medium. Mycoplasma will be transferred from animal cells or from an external contaminated culture. Media contents like sera, peptone etc are also can be a source of mycoplasma.

Chapter 17

Animal Cell Culture

Different cell line collections are available from different cell banks. One can get a suitable cell line collection and multiply in a suitable medium in the laboratory for research purpose. Large scale multiplication of cell lines can be done using a fermentor. In India 'National facility for animal tissue and cell culture' is a centre for cell line collection. The different categories of cell line collection are

1. CRL-Cell repository lines-Human skin fibroblast, cancer cell lines, hybridomas etc. Microbial contamination and species are verified.
2. HB-Hybridomas
3. TIB-Tumor Immunology cell Bank
4. HTB-Human Tumor cell Bank
5. CCL-Certified Cell Lines- Characterized and verified cell lines.

1. Tissue Collection and Culture
Primary Explant Culturing

The tissue can be chopped finely, rinse with sterile media 4 to5 times and small pieces are added to culture flasks or Petri dish with the sterile medium.

Medium-HAM'S F-12-with 10% foetal calf serum, 100ml of BSS (Basal Salt solution), Petridishes-10mm, Forceps, scalpels, pipettes, centrifuge tube, culture flasks etc. All should be sterile and pre incubated to the culturing temperature, and kept in sterile laminar flow.

Transfer the tissue into the sterile BSS in a Petri dish and rinse 4 to 5 times. Transfer to another new Petri dish and remove unwanted tissues. Then transfer to sterile centrifuge tube with the help of sterile pipettes connected to vacuum pump. Resuspend in fresh sterile BSS. Allow to settle. Centrifuge at 600 rpm for 5 min. Remove supernatant. Repeat this 3 times and transfer the tissue to the culture flask with 1ml medium per 25cm^2 culture tissue surface. Incubate at 37^{0}C for 18 to 24 hrs. Media may be added for 3 to 5 days and may be changed weekly till out growth is observed.

2. Trypsinization of Cells

The tissue are chopped and mixed with trypsin for dissociation of the cells.

Materials : BSS Medium-50ml, Trypsin-2.5% in 1N PBS -200ml, DMEM sterile with fetal calf serum, Sterile culture flasks-5 nos, Sterile Petri plates, pipettes, centrifuge tubes, Haemocytometer etc. All items should be sterile.

Procedure: Collect the tissue, Chop it and wash in the basal medium.

Transfer the collected tissue to sterile BSS in a 10mm sterile Petri dish. Rinse well. Transfer to next dish. Remove unwanted tissue using a sterile scalpel.

Transfer the tissue to a centrifuge tube.

Add 5 ml of the sterile BSS. Allow to settle. Discard the supernatant.

Wash it in the BSS for 3 times.

Transfer the tissue to new flask. Add 90ml of sterile PBS.

Add 10 ml of 2.5 % trypsin.

Cap the flask. Mix well with the help of magnetic stirrer kept at temperature 37^{0}C.

Stir at 100rpm for 30 min.

Collect the disaggregated cells and repeat trypsin treatment as before.

Collect the disaggregated cells and resuspend cells in 10 ml of culture medium with serum.

Count the cells in a haemocytometer.

Dilute appropriately to get a cell concentration of 1 x 10^6 per ml.

Note: Some cell need cold trypsinization to avoid cell damage. For this tissue has to be chopped and kept in trypsin at 4^{0}C for 10 to 18 hrs.

3. Cell Separation by Density Gradient Method

Cell separation can be done using percoll gradient. Percoll at different densities are layered and centrifuged with the sample. Collect the fractions, dilute and use for culturing.

Requirements: Growth medium, Percoll 20%, 1N PBS, Trypsin 0.25%, Syringe, culture, microtitre plates, centrifuge, haemocytometer.

Procedure: Adjust the density of the percoll to 1.1 gm/ml and osmolarity to 290 mOsm /Kg.

Mix the percoll with the culture media to different density range, for example- 20%, 45%, and 90%.

Layer the percoll one over the other. Centrifuge at 5000rpm for 10 min. Collect trypsinized cell in the medium and place it over the percoll gradient prepared. Centrifuge at 5000 rpm for 20 min. Cells will be separated in different fractions. Collect the fractions as 1ml in a micro titer plate or 24 well plates. Add equal volume of the medium in each well. Mix well. Incubate for 24 hrs and remove percoll and add medium for growth.

Note: Depending on the cells to be collected, the density of the percoll, time and speed of centrifugation etc varies. Ficoll also can be used for density gradient.

4. Low Speed Separation of Cells

Materials

Centrifuge tubes, Clinical centrifuge with fixed angle rotor, Phase contrast microscope, Whole Blood, Phosphate Buffered Saline (PBS) + 0.1% (w/v) Bovine Serum Albumin (BSA)

Procedure

Collect at least 5.0 ml of blood into a sterile storage vial containing EDTA as an anticoagulant. Add 5.0 ml of EDTA anti-coagulated whole blood to a centrifuge tube. Mix gently by inversion. Centrifuge at 2,000 x g for 10 minutes at room temperature with a fixed angle rotor. Decelerate the centrifuge slowly (do not use a brake). Keeping the centrifuge tube upright, insert a micropipette into the tube and push until the band (mononuclear cells) is displaced into the micropipette. Do not collect more than 2.5 ml into the pipette. After the centrifugation of whole blood, there should be an opalescent compact band just below the meniscus of the tube. This band contains the mononuclear cells (lymphocytes and monocytes) and platelets. Below this band is the plasma, while the erythrocytes and polymorphonuclear cells will form a dark band at the bottom of the tube. Transfer the contents within the tube by inverting it into a 15 ml conical centrifuge tube. Rinse the inside of the tube with 2.5 ml of PBS and transfer the washing to the conical centrifuge tube. Wash the mononuclear cells in the conical tube by adding 5 ml of PBS containing 0.1% BSA for a total volume of 10 ml and mix by inversion. Centrifuge at 300 x g for 10 minutes at room temperature. Resuspend the mononuclear cells in 5.0 ml of PBS. Make a simple wet mount of the suspended cells and examine the purity of the separation using a phase contrast microscope.

5. Isolation of Lymphocyte

Lymphocyte can be isolated from blood. It is present in blood, peritoneal exudates or lymphoid organs mixed with other cells. Various separation procedures can be used to obtain pure population of lymphocytes.

Lymphocyte can be separated using differential migration, following centrifugation results in the formation of several layers of cells. Mononuclear cells (lymphocytes and monocytes) and platelets will be in the interphase band due to their density. In the pellet granulocytes and erythrocytes will be present. Lymphocyte will be in the plasma layer. It can further be purified using several washing with isotonic phosphate buffered saline.

Materials : *Separation reagent*: Sucrose-5.7 g/dl, sodium diatrizoate-9g/dl.Osmolarity 300mOsm, pH-8.2,

Sterile centrifuge tubes, pipettes, centrifuge

Blood samples : Should be collected freshly in heparin or EDTA coated vials. It should be processed within 2-3 hrs. Use a 1:2 dilution of the blood sample with sterile normal saline or PBS. It should be isotonic with the blood sample.

Procedure : Take 2.5 ml of the separation reagent in sterile centrifuge tube. Overlay with 7.5 ml of the diluted blood. Do not mix. Centrifuge at 1600 rpm for 20 minutes. A band/layer will be formed and it contains the lymphocytes. It can be carefully separated without mixing. It can further wash with isotonic PBS for pure separation. Layer PBS on the lymphocyte band gently and centrifuge again for 10 min. Culture media also can be used for isolation instead of PBS. Observe under microscope. Count the number through a haemocytometer after staining with tryphan blue. It stains viable cells.

6. Cryo Preservation

Cells can be stored long time using controlled freezing and storage.

Collect the cells in log phase and prepare a high density suspension in a suitable preservative and freeze slowly.

Materials : Cells which are to be stored (Spleen cells, embryos etc.), growth medium, preservative such as glycerol, freezer

Procedure : Collect the cells in the medium. Add glycerol 10%. Freeze it using a programmed freezer as 1^0 C/minutes. Collect, label and store the vials in liquid nitrogen.

7. Isolation of Spleen Cells

Aim : To isolate spleen cells of mouse BALB/c

Principle : Spleen is a secondary lymphoid organ in the abdominal cavity. Lymph nodes trap the antigen from local tissue while spleen filter and trap the blood born antigen.

Thus it can respond to the infections. Unlike the lymph nodes spleen does not have lymphatic vessels. Instead blood bovine antigen and lymphocytes are carried into the spleen through the spleen cavity .Radioactivity studies shows that more recirculation of lymphocytes passes through spleen than all lymph nodes combined.

The spleen is surrounded by a capsule that extends a number of projections into the interior to form a compartmentalized structure. The compartments are of two types, the red and white pulps which are separated by a diffused marginal zone.

The red bulb of the spleen consist a of a network of sinusoids populated by macrophage and numerous red blood cells .It is the site where old and defective red blood cells are destroyed and removed. Many macrophages within the red blood cells are iron pigments from degraded hemoglobin. The spleenoic white pulp surrounds the branches of the spleen cavity forming a perianterior lymphoid sheets (PALS) populated mainly by T lymphocytes. The marginal zone located peripheral to the PALS is rich in B cells organized into the primary lymphoid particles.

Blood borne antigens and lymphocytes enter the spleen through spleen artery which empties into the marginal zone and antigen is trapped by dendrite cells which carry it into PALAPs.

Materials

BALB/c mice, dissection set, chloroform, cotton, laminar flow and RBC lysis buffer.

Procedure

A healthy BALB/c mouse has to be selected. Take the mouse and wipe with 70% alcohol. Put the mice in jar with chloroform. Open the peritoneal cavity and remove the spleen and place in 5 ml of ice cold PBS in a Petri dish. Flush the cells and collect it in PBS. Centrifuge at 1000 rpm for 10min to obtain the pellet. Resuspend in RBC lysis buffer kept in ice and centrifuge at 1000 rpm for 10 min. Resuspend in ice cold PBS and centrifuge again for 10 min at 1000ppm.Then the cells were suspended in RPMI-1640.Few drops are taken and observe under microscope. Count the number of cells per ml by using a haemo-cytometer.

Result : Spleen cells were collected and the number of cell per ml obtained is:

8. Isolation of Peritoneal Macrophage

Aim : To isolate peritoneal macrophage from mice BALB/c

Principle: The mononuclear phagocyte system has two main functions prepared by two different types of bone marrow derived cells. Phagocyte macrophages remove the antigen. Antigen presenting cells can take up, process and present antigens to T cells.

The mononuclear phagocyte system includes blood monocytes and phagocytes and are resistant in tissue or fixed to the endothelial layer of blood capillaries. In the liver macrophages are known as Kupffer cells while in kidney it is known as inter glomerular messengial cells. Alveolar and peritoneal macrophages are wandering macrophages.

The phagocyte tissue macrophages together with endothelial cells which are previously constituted are known as reticular endothelial system. Phagocytic macrophages are found in many organs and can be revealed by intravenous injection of carbon particles which become localized in the phagocytes. Myeloid progenitors in the bone marrow differentiate into pro monocytes and then into blood monocytes. Cells from this circulating pool migrate through the blood vessel walls into the various organs to become macrophages. The human blood monocyte is larger (10-18 micro diameter) relative to the lymphocytes. It has a horse shoe shaped nucleus and contains azourophillic granules. The monocytes posses diffused membrane, a well developed golgi complex and many inter cytoplasm lysosomes. These lysosomes contain peroxidase and several acid hydroxylases which are important for intracellular killing of microorganism.

Monocyte/macrophages adhere strongly to the glass and plastic surfaces and actively engulf the organism or even tumor cells *in vitro*. Adherence and ingestion by monocytes occurs when the cells bind the microorganism by specialized receptors .The receptors may bind to certain carbohydrates of the microbial cell wall or to the IgG and complement with which the microorganism has become coated. Elicitor solution when injected into the peritoneum causes rapid increase in peritoneal macrophage population.

Materials : Dissection box, Needle, 70 %ethanol, Laminar air flow, 21g needle, phosphate buffer saline, RPMI 1640 medium, 19 g needle, 10 ml syringe.

Elicitor solution-Tryptone-1.7%,Peptone-0.3%,Glucose-0.6%,NaCl-0.25%,Sodium thioglycollate-0.05%,Agar-0.05%,double distilled water-100ml,Sterilize.

Procedure : Two healthy BALB/c mice are selected. 1 ml of the elicitor solution is injected on the lower right side of the peritoneal cavity with sterile 21 gauge needle. Then keep the mice in a cage for 5 to 6 days for macrophage development. The animal is killed by keeping in a chamber containing chloroform. Take the mice and keep on a dissecting tray and wipe with 70% alcohol. Remove the outer skin and inject 5-8 ml of ice cold PBS to the peritoneal cavity with a 21 g needle. The cavity wall is messaged gently to release the macrophages to the fluid. The peritoneal fluid is collected with a 21 g sterile needle. It is immediately kept in cool conditions to avoid adhering to the wall of the container. The macrophages collected are washed 2-3 times with sterile RPMI medium and transferred to the culture flask. The cells are incubated in a CO_2 incubator at 37^0 C for 3 hrs. Few drops are collected and observed under microscope. Take the count using a haemocytometer.

9. Isolation of Blood DNA

Aim : To isolate Blood DNA.

Materials required : Eppendorf tubes 2ml -10nos, tips, centrifuge,

Buffer-1N PBS, Na phosphate -10 Mm,KCL-3Mm,NaCl-140mM,Na Phosphate-10Mm,pH-7.2.NaCl 2m,NaOH-0.1M.

Procedure : Take 250 μl of blood in a sterile eppendorf tube which is anti coagulated with EDTA. Add 1 ml of sodium phosphate0.01M and pH7.2. Mix and centrifuge at 10,000 rpm for 1min.

Repeat the same (step 2).

Resuspend the pellet in 1ml of PBS [NaCl 140 mm, KCl-3mM, $Na_2(PO_4)_2$-10mM, pH-7.2]. Centrifuge at 10,000rpm for 1min.

Dissolve the pellet in 10 μl of 2mM NaCl, 0.1M NaOH

Incubate at 95 ^{0}C for 2min.

Centrifuge at 10,000rpm for 10 min. DNA will be in the supernatant.

The DNA can be precipitated by adding ice cold isopropanol. Gently mix. Centrifuge at 10,000 rpm for 10 min. Collect the pellet.

Wash it with 70% alcohol by repeated precipitation and centrifugation with 70% alcohol.

Dissolve in TE buffer at pH 7.8 and can be stored.

Check the presence of DNA by adding 2 µl of the sample on an agarose gel with ethidium bromide in a petri plate and subject it to air dry. Observe in a gel doc. Florescent spot is an indication of the presence of DNA.

DNA isolated can be used for electrophoresis using agarose gel.

10. Differential Count of Leucocytes

Aim : To get the differential count of leukocyte of peripheral blood.

Principle : The bone marrow produces the red blood cells, thrombocytes, neutrophils, esinophils, basophils and serves as the site for the production of lymphocytes, monocytes, and plasmocytes.

Erythrocytes : Mature erythrocytes are 6-8 µm in diameter and 1.5 µm to 2.5 µm in thickness and a surface area of 130 to 150 square µm. Red cells stain to a reddish color. I t is biconcave in shape. Cells appear light in color in the centre than the periphery.

Some of the cells may have cup shaped or spherical shaped cells.

Neutrophills : It is a granulocyte with nucleus separated into lobes which are connected by tiny filamentous / strands. The cytoplasm stains pink .The size is twice that of red cells.

Esinophills : Esinophills are of same size with neutrophills and usually have two lobed nucleus. It is spherical and uniform in size usually distributed and stain bright red.

Basophills : Basophiles have prominent dark meta chromatic granules which vary in size shape and number. Mature basophills are comparable in size to neutrophils and constitute less than 1% of the nucleated cells of the bone marrow or peripheral blood.

Lymphocytes : The size of the lymphocytes varies widely. Majority of the lymphocytes are slightly larger than red cells. The nucleus is oval in shape. The nucleus stains darkly and plaque areas of condensed chromatin are separated by lighter areas.

Monocytes : The monocytes vary in size from 15-20 μm and generally longer than nuetrophills. Oval in shape and multiple protrusions are present on the surface. The nucleus is usually accentric and may be round, oval or bilobed. The cytoplasm is bluish grey color and that is quite distinctive.

Staining : Leishman stain is used generally for differential staining. Methylene blue and its derivatives are widely used .Being basic dye methylene blue stains the nucleus of all cells. The staining properties of the cytoplasmic granules differ and can be classified as acidophilic (esinophills), basophilic (basophills), and neutrophillic (neutrophills).E sinophills stain violet color to nucleus and reddish orange to cytoplasmic granules. Basophills show blackish blue granules overlapping the nucleus. Nuetrophills have fine pink granules. Lymphocytes have large nucleus.

Materials : Clean glass slide, sterile needle, Leshman's stain-0.15 gm of Leishman's stain powder in 100 ml of methanol.

Phosphate buffer : 1.21 gm of K_2HPO_4 0.3 gm of KH_2PO_4 and 8.0 gm of NaCl in 1000 ml of distilled water, Immersion oil.

Procedure : The finger tip is pricked with sterile needle. The first drop blood is wiped and second drop has to be collected on a clean slide. Make a smear on the slide with the help of another slide dragging on it. It is then allowed to air dry. Add Leishman stain and allow reacting for three minutes. The stain is then diluted with PBS. Mixed well and drained off. Washed and observed under microscope. The number of each type of cells is counted.

Red cells were all stained pink and will not be overlapped.

Result : Number of lymphocytes

Number of Neutrophills

Number of monocytes

Number of esinophills

Number of basophills

Note : The normal percentages are

Lymphocytes-20-40%

Neutrophills-50-70%

Monocytes - 1-6%

Esinophills-1-5%

Basophills-0-1%.

11. Slide Gel Electrophoresis

Aim: To separate serum protein by slide gel electrophoresis.

Principle: Globulins are present in higher concentration (60 to 80%) in the blood. Fibrinogens are essential for clotting of the blood. Serum globulins are generally antibodies. Globulins exists as $\alpha1$, $\alpha2$, β, γ globulins.

Materials

Tris glycine buffer: Tris base-3.0 gm

Glycine-14.4 gm

Distilled water-1000 ml

pH 8.4

Commassie brilliant blue stain

Agarose - 0.5 % in tris - glycine buffer – 100 ml

Conical flask, glass rod, beaker, slides, Electrophoresis unit and power pack.

Procedure

Prepare 0.5 %agarose in tris glycine buffer. Place it on a clean slide, allow settling. Four plates have to be prepared and place on the platform of the gel tank. Keep a paper wick so that slides are connected to the buffer in the gel tank. Cut a small filter paper and dip on the serum sample and placed on the slide towards the cathode side. Close the apparatus and run the gel at 120volt for 3 hrs. After running for 3 hrs, remove the slide and place in absolute alcohol for 10 min for denaturing the protein. Then the slides has to be immersed in commassie brilliant blue solution for 10 min. Wash slides with 40% acetic acid for de staining.

Result: Bands are obtained showing $\alpha 1$, $\alpha 2$, β, γ globulins and albumins.

12. Haemagglutination Assay

Aim: Haemaglutination of red blood cells for antibody measurement.

Principle: This is generally used for the detection and measurement of antigens. When RBC's are added to antigens in serial dilution, agglutination occurs and the highest dilution that produces haemaglutination and it is referred to as the haemaglutination titer. This can be carried out in a series of test tubes or in plate with wells. Red blood cells which are not in agglutinated forms like a button at the bottom and agglutinated cells are seen spread into a shield like pattern.

Materials: Goat RBC, PBS, BSA in saline, Microtitre plates, micropipettes, centrifuge tubes, Needle and syringe.

Procedure: Immunize the mice with goat RBCs 2 days prior to the blood collection. Collect the blood by bleeding the tail end portion. Collect 500 micro liters of blood. Centrifuge at 5000 rpm for 10 min. and serum is collected. Heat the sample at 60^{0}C for 30 min to inactivate the complement. The serum is then diluted in the ratio 1:1,1:2,1:4,1:8 with PBS.50 micro liters of serum is then added to each well .One well of serum is kept as control.50 micro liters of 0.1 %BSA is then added to 10 micro liters of goat RBC in each well.

Incubate the plate at 37^0 C for 1 hr.

Plates are observed for agglutination and titer was noted.

13. ELISA

Aim: To estimate the IgG concentration in the blood.

Principle: Enzyme Linked Immuno Sorbant Assay (ELISA) is highly specific and reproducible method for measuring the IgG in the blood cells. This was described by Enguall and Pertmann in the year 1972.

In this assay the antibody is adsorbed on the well surface of the plate .After blocking the unbound sites incubated with another antibody.

Materials: 96 well micro titer plates/ELISA plates, Bicarbonate buffer (0.1 M)-84 gm of Na HCO_3 in 1000ml of distilled water. (Dilute 1 to 10 ml)

PBS: 1.21 gm of K_2HPO_4, 0.34 gm of KH_2PO_4, 8 gm of NaCl. pH 7.2

Blocking buffer- 1%BSA in PBS with 0.1% tween 20,

Washing buffer-PBS with 0.05 % Tween 20,

Substrate buffer-Citrate phosphate buffer-citric acid-7.3 gm,

Na_2HPO_4-11.86gm, 100ml distilled water.

O-phenyl di amino di hydrochloride-substrate

Horse raddish peroxidase,

Hydrogen peroxide,

Antibodies.

Procedure : Coat the ELISA plate with the monoclonal antibodies hg 11 with 150 micro liters of protein sample having a concentration of 10 micrograms/ml and incubate for 4 hrs. Wash with washing buffer bock with blocking buffer at 37^{0}C for 2 hrs. Wash with PBS-TWEEN and 50 micro liters of sample and control are added in duplicate in separate wells. Incubate at 37^{0}C for 3-4 hrs. Wash with PBS tween .Add 150 micro liters of HRP labeled goat antiserum and incubate at 37^{0}C for 24 hrs. Wash with PBS and add substrate 150 micro liters in each well. Yellow color will be developed in positive samples. Stop the reaction with 1N H_2SO_4.

Read the absorbance at 490 nm.

Result : The positive reaction gives the yellow color and indicates the presence of antibody in the test sample.

Note : Enzyme reactions should be carried out in dark in order to avoid false results.

14. Western Blotting

Aim: To perform western blot of the given serum sample

Principle: The main step in the western blotting is the transfer of protein bands from the poly acryl amide gel to the membrane of nitrocellulose paper .It binds strongly and available for specific protein ligand binding reactions. Mostly antibodies are used to detect antigens. In most cases probe itself is radioactively labeled and used for the detection.

Materials: 1.5 M Tris base-90.82 gm in 500ml of distilled water.pH8.8.

0.5M Tris-60.55g of Tris in 500 ml of distilled water. pH 6.8., 10% SDS,

30%ACRYL AMIDE with 0.8% BIS ACRYLAMIDE, (100ML) (*Danger neuro toxins*)

3X sample buffer-SDS-0.75gM,Glycerol-1.875ml,250 mM EDTA-0.3 ml,0.5 M Tris HCl ph6.8-7.5 ml,0.5% bromophenol blue-0.3 ml,EDTA-0.093034gm/ml,Final volume-10ml.

5X running buffer-Tris BASE-30 gm, Glycine-144gm, 10% SDS -Make up to 2liters.

DYE

Commassie brilliant blue-1.0 gm, methanol-250 ml, Acetic acid-37.5 ml, Distilled water-162.5 ml

Gel destain solution

Methanol-25 ml, Acetic acid-37.5 ml, Distilled water-437.5 ml

Ammonium persulphate-10 gm in 100ml

TEMED - 100 ml

Procedures

Clean the glass plates of the electrophoresis unit. Wipe with isopropanol and clamp two plates with spacers between them. Pour the separating gel in between the plates and allow to solidify. Pour the stacking gel above the separating gel and place the comb immediately. Allow to solidify. Insert the glass plate with the gel in the electrophoresis unit. Fill the electrophoresis tank with running buffer and load the sample into the wells. Run the apparatus at10 volt for 3 hrs. After running, the gel is removed and placed in the transfer buffer. Cut the PVDF membrane exactly to the size of the gel. First take the cassette and gel. Wet the sponge. Place what man no 1 filter paper in two layers between the sponge and gel. Finally place another sponge over it and close the cassette. Keep in the electrophoresis unit and run for 2 hrs at 200 Milli amperes and 0.3 Milli volt. Carefully remove the gel and stain with commassie brilliant blue for 30 min. Then destain it in the destaining solution for 5 min .See the appearance of the blue band. Wash the PVDF paper with solution PBS having 0.1 % tween 20 and transfer to the antibody binding buffer (5% milk powder in TBS, 0.1 %Tween 20, 5 micro liters of secondary conjugated with horse radish peroxidase). Place it on a rocker for shaking overnight. Then transfer the PVDF paper to blocking solution. Wash the paper with TBS Tween 20, add 150 micro liters of OPD substrate and incubate for 30 min. Wash again with PBS-Tween 20.

Result: Bands are absent on the gel indicating that it has been transferred to the PVDF and the transferred bands are observed on the membrane.

15. WIDAL Test

Aim : To diagnose the enteric fever in the given serum sample.

Principle : Salmonella typhi, salmonella paratyphi are the causative organisms of enteric fever commonly known as typhoid fever. The test is based on the antibody antigen reaction which results in agglutination .Antigen O is present on the surface of all salmonella species and Antigen M present is different for different salmonella species.

Materials : Serum sample, Test tubes, normal saline, kit.

Procedure

5 clean test tubes are arranged serially.7.0 ml of normal saline is taken in first test tube and 3.5 ml to 2,3,4 and 5th tubes. 0.5 ml of serum is added to first tube and mixed well.

Then from this first test tube 3.5 ml is transferred to second tube and 3.5 ml from the second to third test tube. Then 3.5 ml is transferred from third to fourth test tube and finally3.5 ml is transferred from fourth to fifth test tube to get a serial dilution of 1:15, 1:30, 1: 60, 1: 120 and 1: 240 respectively.

24 test tubes are arranged in 6 x 4 rows

0.5 ml of dilution from tube 1 is transferred to each tube in row 1.Similarly 0.5 ml form tube no 2 to second row. Repeat this process up to the transfer of 0.5 ml from tube no.5 to the fifth row. 0.5 ml of normal saline is added to each tube. One raw will serve as control and in other rows add 0.5 ml of the S.tyhi o, H, A (H), and B (H). Mix well and incubate at 37^0 C over night.

Observation : Observe the agglutination in each dilution with naked eye. Find out the highest dilution in which the agglutination is present. For H antigens the agglutination pattern is cotton wood type and for o-antigen the pattern is fine granular and granular mat for a-antigens.

16. Identification of Y-chromosomes in Cattle using PCR.

Aim: To identify Y chromosome specific gene in cattle for sex determination.

Principle: Y chromosome specific gene can be identified using specific primers and can use for sex determination.

For example: BRY 4-5′

CTCAGCAAAGCACACCAGAC-3′

5′ GAACTTTCAAGCAGCTGAGGC-3′

Procedure: Collect blood from cattle in heparin coated vials. Isolate genomic DNA from cattle blood as explained in experiment 7.2.8

PCR-Volume of the total reaction mixture is 25 μl

PCR buffer: 1X- (50mM KCl, 10Mm Tris- HCl, pH-9.0, 2.5mM $MgCl_2$, and 1%Tritron X-100).

200 μM of each dNTP's

0.5U of Taq polymerase

20 μM of each primer. Overlay the reaction mixture with 100 μl of light mineral oil.

PCR cycles are given below

I- In the first cycle the denaturation is at 97^0 C for 3 min.

II- 40 cycles : 94^0C for 30 sec, 58^0C for 45 sec and 72^0C for 1 min. After completion of 40 cycles, keep the reaction mixture at 72^0C for 5min and then cool to room temperature.

Run the PCR product on agarose gel (2% with 0.5 μg/ml ethidium bromide).Visualize the bands using gel documentation system. If band size comparable with 300bp is obtained, it can be assigned as male.

Note : Different kits are available for specific PCR amplifications. It can be performed for demonstrations. For specific amplification of a new gene, standardization of PCR conditions is required.

Chapter 18

Embryonic Stem (ES) Cells

Introduction

Embryonic stem (ES) cell- mediated alterations of the mouse genome have revolutionized the genetic approaches that can be employed in human biology and medical research. Such genome modifications have provided invaluable information in all fields of the life sciences concerned with normal and diseased biological function. ES cells can develop into any type of the tissue of the body. So the ES cells can be used for tissue transplantation treatments. It is an alternative method of treatment in degenerative diseases. Chord blood is a good source of stem cells and can be preserved for a possible use in later life of an individual if required.

Embryonic cells are derived from preimplatation embryos, specifically from inner cell mass (ICM) of the blastocyst stage. After establishment, these cells can retain the potential of one of the components of the ICM, the primitive endoderm. ES cells can differentiate efficiently into germ cells. ES Cells are good alternative way to create transgenic animals. One of the most powerful transgenic technologies made possible by ES cells is gene targeting based on homologous recombination.

The new generation of tools combines homologous and site-specific recombination so efficiently to create new knock out animal ES cell models useful for gene therapy, tissue transplantation research.

1. Maintaining Embryonic Stem (ES) Cells

There are several good germline-compatible ES cell lines currently available from different firms/cell banks.

Common media used for ES cell preimplantation embryo culture

a. To culture ES cells

DMEM + (with high glucose content)

DMEM + (2mM glutamine)

DMEM + (1mM sodium pyruvate)

DMEM + (15% fetal calf serum)

DMEM + (1000U/ml LIF)-Necessary if ES cells are cultured without feeder cells.

DMEM + (PBS with out Mg and Ca)-For washing ES cells.

M16 MEDIUM- For culturing embryos.

2. Establishing ES cells

1. Recover embryos from female mice ovaries after 3.5 days postcoitus. The embryo preferably should be in blastocyst stage. Place one embryo in one well on feeder cells present in a 24 well plates with ES cell culture medium. Incubate in a CO_2 incubator.

2. After attachment (usually two days) of the embryos, replace half of the medium daily with prewarmed at 37^0C fresh medium for 3 to 4 days. During this period ICM of the blastocyst increases.

3. Change the medium to PBS. Disaggregate the blastocyst out growth by transferring to 96 well plates with 50 micro liters of Trypsin .Keep for 20 min. and transfer to fresh plates with feeder cell lines.

4. Replace half of the medium with prewarmed fresh medium for 3-6 days. At this point ES cell-like colonies will appear.

5. ES cell like colonies can be disaggregated and repeat the steps 3 and 4 till get enough ES cells.

Note: For some ES cells, feeder cells are not required but gelatinized culture plate or flask can be used for the culturing of ES cells. The feeder cells provide nutrients, growth factors and matrix constituents that enable the cloned cells to survive more readily.

Preparation of feeder layers: Collect 13 days old mouse embryo fibroblast cultures. Trypsinize as in previous experiments, reseed as 10^5 cells/ml. Add mitomycin C in the concentration 2 micrograms/10^6 cells (25micrograms/ml) .Keep overnight. Change the medium after treatment. Again trypsinize and reseed in fresh medium. Incubate for 24 hrs and can be used as a feeder layer. The feeder layer / cells will be viable for2 to 3 weeks. Other cells or homologous cell lines also can be used as feeder cells.

3. Freezing of ES Cells

Aspirate ES cell medium from the plates. Wash by replacing with 10 ml of PBS .Replace PBS with 2ml of Trypsin and incubate for 5 min at 37^0C.Stop the action of Trypsin by adding ES cell medium and transfer the cells to 15 ml centrifuge tubes. Spin at 1000 g for 5 min. Aspirate supernatant .Suspend the pellet in 2ml of ES cell medium gently. Distribute the cell suspension into 1.5 ml cryovials and add 0.5 ml of 2x freezing medium. Place the cryovials at -70^{0}C for 10 hrs and transfer to liquid nitrogen for long term storage. DMSO can be used as cryoprotectant. Glycerol can also be used as cryoprotectant. When exposed to light glycerol will be toxic to the cells. So for long term preservation, glycerol is not recommended.

4. Thawing of Frozen ES Cells

Place the frozen vials in a 37^{0}C water bath. After the ice is melted, immediately transfer the contents to a centrifuge tube and slowly add 4 ml of fresh prewarmed medium. Spin at 1000g for 5 min. Aspirate the supernatant and gently resuspend in 5 -8 ml ES cell medium. Now this can be used for the culture purposes.

5. Production of Transgenic Using ES Cells

An exogenous DNA can be introduced to the ES cell by electroporation. The transformed ES Cell can be identified using a drug resistance gene implemented into endogenous DNA and give the specific drug resistance to the cells that carries the vector. The most commonly used selectable markers are neomycin phosphotransferase [(neo), 7-9 days for selection], puromycin (3-4 days for selection), and hygromycin (15-20 days for selection).

Various steps in creating the transgenic

1. Collect ES cells with concentration 10^6 to 10^7 cells/ml in ES cell media
2. Prepare DNA vector, linearize, purify by ethanol precipitation, and make into a concentration of 1ìg/ìl in a sterile mixture of 10 mM tris and 1 mM EDTA or in sterile double distilled water.
3. Trypsinize the cells as before.

4. Count the cells and adjust the cell concentration to 7 x 10^6 cells/ml in PBS.
5. Add 10-40μl of DNA and 800μl of ES cell suspension into 0.8 ml electroporation cuvettes.
6. Apply an electric pulse of 500μF at 250 V through the Cuvette.
7. Keep the Cuvette on ice for 20-30 min.
8. Transfer the cells from the Cuvette to the ES Cell medium in a culture flasks or plates.
9. Allow to grow for 3 days in CO_2 incubator.
10. Add selective drug to the medium and select the surviving live cells. Use the lowest concentration of the drug to kill the non resistant cells.

6. MTT Assay for Cell Viability

Aim: To check the cell viability using MTT

Principle: MTT is a yellow water soluble tetrazolium dye and can be reduced by live cells to a purple colored insoluble product. MTT is reduced by the oxidative metabolism of the cells to give formazon. This can be observed by naked eye and measured spectrophotometrically. The number of viable cells can be counted from the OD value.

Materials: MTT-5mg/ml in PBS, pH 7.4

Stop solution: 0.2 M HCl in isopropanol.

RPM1 medium

Culture plates

Centrifuge tubes

Procedure: Mononucleated cells are collected from human blood and dilute 1:1 with PM1 medium. (Cell density-10^6 cells /ml).100 µl of the suspension was added to each well. Add equal volume of the media in each well. Incubate over night for growth. Add 15 µl of MTT in each well and incubate for 2 hrs and observe for cell viability.

Result: Color change indicates the presence of viable cells (purple colour).

Section 8

MOLECULAR BIOLOGY

Chapter 19

General Molecular Biology Procedures

1. Isolation and Purification of DNA from *E.coli*

Aim : To isolate and purify DNA from E. coli culture

Principle: In order to isolate DNA from the nucleus, cell wall has to be broken and removed. This can be achieved by adding detergents such as SDS. Cell contains protein and this can be removed by adding phenol which deprotienize and removes proteins to inter phase. Nucleic acid will remain in the aqueous layer. On centrifugation aqueous layer with nucleic acid will be on the top layer and proteins will be in the inter phase and phenol in the lower organic phase. By adding chloroform isoamyl alcohol, traces of phenol can be removed. The DNA from the aqueous layer can be precipitated by iso propanol or 70%ethanol.

Requirements

LB medium

Extraction buffer/lysis buffer:-100ml

10 mM Tris chloride-pH-8.0, EDTA-0.5 M, NaCl-10Mm,1%SDS

Sodium acetate: 3M-pH 5.2-100ml

70% ethanol.

Equilibrated phenol-100ml (phenol is acidic, basic Tris (pH-11.0) is used to equilibrate phenol)

Chloroform: isoamyl alcohol-1:1

E.coli-Overnight grown cultures, O.D at 600nm should be 0.8

Extraction

Collect 5 ml of the overnight grown E.*coli* culture.

Centrifuge at 5000 rpm for 10 min. Decant the supernatant and collect the pellet. Wash with 5ml of ice cold phosphate buffer (pH 7.6). Mix well. Again centrifuge at 5000 rpm at 10 min. Discard the supernatant and pellet is resuspended in lysis buffer (2.5 ml). Mix well. Keep in ice. Add equal volume of phenol to this mixture and mix gently. Centrifuge at 5000 rpm for 10 min. Collect the supernatant. This is repeated 2-3 times. Add equal volume of CIA to the supernatant. Mix well. Centrifuge for 5000rpm for 10 min. Collect supernatant. Add equal volume of 70%ethanol to this. Mix gently. DNA will be precipitated. Repeat this step 2-3 times. Ethanol is evaporated and precipitate is dissolved in TE of pH-8.0.

Purity of the DNA is checked by taking OD at 260 and 280nm. It is 1.8 for pure DNA. Less than 1.8 is an indication of protein contamination in the DNA preparation.

Note: Genomic DNA can be isolated using spooling method. In this method 10 % SDS is added to suspended E coli pellet to get a 1% SDS. Mix gently and incubate at 65^0C for 15 min (To deactivate DNAase).Cool to room temperature and add isopropanol. DNA will be precipitated.

2. Isolation of Plasmid DNA

Aim: To isolate plasmid DNA

Principle: The isolation of DNA is based on the cell wall breakage and release of DNA and separation of the DNA from other macromolecules like proteins.

Materials

Solution 1: Tris HCl-25Mm-pH-8.0 + 50 mM glucose + lysozyme - 100mg-10ml

Solution 2: NaOH-0.2 M+ 1% SDS-100ml.

Solution 3: Sodium acetate-2M-100ml

Solution 4: Tris HCl-50mM+sodium acetate-100mM-100ml

Solution 5 : Overnight grown E.coli strain with plasmid.

Procedure : Centrifuge 10 ml the overnight grown culture of E.coli at 3000rpm for 10min.Collect the pellet. Resuspend in 1000 µl of solution 1.Keep in ice for 30 min. Add 1000 µl of solution 2. Mix and keep on ice for 5 min. Add 1500 µl of solution 3.Mix and incubate for 60 min in ice. Centrifuge at 5000 rpm for 5min.Collect supernatant and add double volume of ice cold ethanol. DNA will be precipitated. Centrifuge and collect pellet. Wash with 70% ethanol. Dissolve in TE buffer and store.

3. Gel Electrophoresis of DNA

Aim: Electrophoresis of DNA.

Principle: Agarose gel separates the DNA fragments according to their molecular size .It is an efficient method for separating and purifying the DNA fragments .Under the influence of electric field the DNA fragments migrates according to their molecular size The negatively charged phosphate groups present on the backbone of the DNA imparts negative charge to DNA, which enables the DNA fragment to migrate towards anode. The rate of migration is thus inversely proportional to the molecular weight of DNA.

Agarose is a neutral poly saccharide purified from agar of the red algae. It is soluble in water. Agarose have uniform pore size while migrating DNA has different size and electrophoretic mobility of DNA depends on

i) Molecular size of the DNA
ii) Agarose concentration
iii) DNA structure
iv) Applied current-Power pack

Materials Required

Horizontal gel electrophoresis unit,

Agarose,

DNA standard and DNA sample.

TAE buffer L(Tris Acetic acid EDTA)-1X

Tris-10 mM pH-8

EDTA-1Mm

Acetic acid-57.1 ml

Sample buffer-10 x stock

Bromophenol blue-0.25%

Xylene xyanol-0.25%

Glycerol-20%

Ethidium bromide- 0.5µg/ml (2.5 µl in TAE) (Stock 1mg/ml) (Handle with care)

DNA standard markers-(Usually1-10 nano grams in one well)

DNA sample-Lambda DNA digested with Eco RI, Hind III.

Loading Mixture

i) DNA standards are prepared with the composition-DNA ladder-5 µl, TAE-5µl, Dye-2µl-mix well and load in the first well.

ii) DNA sample-5µl, TAE-5µl, Dye-2µl-mix well and load in the next well. Similarly different samples were loaded in the different wells.

Note : Buffer compositions

i) TAE-Tris base-40Mm, EDTA-1Mm, Acetic acid-40Mm.

ii) TBE-Trisbase-89mM, Boric acid-89mM, EDTA-2Mm

iii) TE buffer- Tris-10 mM, EDTA-1Mm, pH-8

Movement in TAE is faster than in TBE buffer. But bands will be sharper in TBE.

Procedure

Electrophoresis apparatus should be cleaned properly, dried and wiped with 70%alcohol.Place it on the casting tray. Arrange the electrophoresis apparatus by leveling it and placing the comb on required height. Seal the sides with cello tape. 50 ml of 1% agarose gel prepared in TAE 1x with ethidium bromide0.5 µg/ml. Pour the agarose with out bubbles. Allow to solidify (20 to 30 min). Remove the combs gently. Pour the running buffer (TAE/TBE-1X pH-8).Load the samples in the wells. Cover the gel tank. The electrophoresis unit is then connected to the power pack .Power supply of 100V is given for 1 hr. After running the front dye just above the bottom, the gel is taken out and observed in gel doc/UV.

Result: Bands are observed.

Note

1) Disposing of ethidium bromide. Make the concentration of ethidium bromide to 0.5 mg/ml. Add one volume of $KMnO_4$ (0.5 M). Mix carefully and add one volume of 2.5 N HCl. Keep at room temperature for several hours. Add one volume of 2.5 N NaOH. Mix carefully and can be discarded.

2) *Loading buffer*: Loading buffer ensures that the samples are dropped to the well completely. It adds color to the sample so that the movement can be seen through naked eye. Bromophenol blue migrates to the anode almost double the speed of xylene cyanol independent of agarose concentration.

3) *Staining of DNA:* Ethidium bromide has a planar structure that can intercalate between the stacked DNA bases. The fixed position of this group and its close proximity to the base makes the dye to bind to DNA. This leads to the display of an increased fluorescent yield than in free solution.

4. Southern Blotting

Aim: To perform southern blot of DNA gel

Principle: The original method for blotting was developed by southern (year 1975) for detecting the fragments on the agarose gel which were complementary to the DNA in the given sequence. In blotting, nucleic acid samples are immobilized on to the solid support such as nitrocellulose membrane or nylon. This blotted nucleic acid is then used for the hybridization of the probe. General blotting techniques are

1) Nucleic acid from the gel.
2) Dot blot
3) Colony and plaque blotting. When blotting used for the DNA it is called southern blotting, for RNA it is northern blotting and for protein it is called western blotting. Smaller fragments transfer faster than larger fragments. So DNA may be cut into smaller pieces for more efficient transfer. The DNA can be denatured by treating with 0.5M NaOH. Nylon membrane binds nucleic acids by electrostatic interaction. The binding capacity of nylon is 400 to 600 μg / cm^2. DNA transfers to solid support by capillary action. Positive pressure and vacuum also can be used for the transfer. Denatured or single stranded DNA fragment may be added to the hybridization buffer. Radio labeled probe also can be used for the hybridization purpose.

Requirements: Electrophoresis gel with DNA,0.25m HCl, Denaturing solution-0.5 M NaOH, Neutralizing solution-1.5 M NaCl.,1M Tris HCl, pH7.0, Positively Charged nylon membrane, What man no 1 filter paper, Glass plates

Procedure: Treat the gel with 0.3M NaOH. Rinse with distilled water and denature with 0.5 M NaOH for 30 min.

Use 0.5 M NaOH as transfer solution. Recover the membrane and place it on the filter paper, dry it and use for the DNA transfer. Cut a what man no 1 filter paper to suitable size and place it on the clean glass plate and wet it with denaturing solution. Keep it on the reservoir for the transfer buffer. Place the gel slab on the filter paper carefully. Nylon membrane is wetted with the transfer buffer placed on the gel slab. Gently press it. Mark the orientation as I. Three more sheets of the filter paper may be placed over the membrane and stack the paper towel on it. Place a suitable weight over it. After 8hrs, observe the above nylon membrane for the DNA transfer using a gel doc or transilluminator. Observe the gel also to make sure that DNA has transferred.

5. Restriction Digestion

Aim: To perform Restriction Digestion of DNA.

Principle: Restriction enzymes can cut the DNA strands at specific sequence to yield smaller DNA fragments.

Materials: Plasmid DNA, Lambda DNA, Restriction enzymes, Buffer, Sterile water, Eppendorf tubes, Water bath etc.

Reaction Buffer Composition

Tris HCl-100 mM (pH-7.4), NaCl-100mM, DTT-10mM, Magnesium chloride-100mM.

Stopping Buffer

Combination of 20% sucrose, 10% glycerol, 10mM EDTA, 1% Bromophenol blue and 1% xylene xyanol.

Procedure: Make the reaction mix as following

I) Plasmid DNA-2µl+Reaction buffer 10x -2 µl sterile water-14µl+EcoRI-2 µl,

Plasmid DNA-2μl+Reaction buffer 10x -2 μl sterile water-14 μl+HindIII-2 μl,

Plasmid DNA-2μl+Reaction buffer 10x -2 μl sterile water-16 μl (Control),

II) λDNA-2μl+Reaction buffer 10x -2 μl sterile water-14 μl+EcoRI-2 μl,

λ DNA-2μl+Reaction buffer 10x -2 μl sterile water-14 μl+ HIND III-2 μl,

λ DNA- 2μl+Reaction buffer 10x -2 μl sterile water-16 μl

Mix well. Incubate at 37^{0}C in a water bath for 90min.Add stopping buffer 5 μl to stop the reaction. Mix well. Load 20 μl of the sample and do electrophoresis with each sample in separate wells.

Observe the bands and note the difference with the control.

Note: Kits are available for isolation and purification of DNA and RNA from different sources, ligation and restriction digestion of DNA.

PCR and cloning kits are useful for demonstration purpose.

APPENDICES

Appendix A

MEDIA COMPOSITION

Agar

Agar-1.5 gm
Melt in 100ml of distilled water
Autoclave.

Agar with Minerals

Agar-1.5 gm
Ammonium phosphate ($NH_4H_2PO_4$-0.1 gm)
Potassium chloride (K Cl)-0.1 gm
Magnesium sulphate ($MgSO_4$)-0.1 gm
Distilled water-100ml
pH-7.2-7.4
Agar- 2% (Optional)
Basal medium
Tryptone-0.2 gm

NaCl-0.5-gm
Di potassiumphosphate-0.03 gm
Glucose-1.0 gm
Bromothymol blue-0.008gm
Distilled water-100ml
pH-7.2-7.4
Agar-2% (Optional)

Blood Agar

Make 100 ml of nutrient agar with 0.8 % NaCl

Distilled water-95 ml

Autoclave. Cool to 60^0C. Add 5 ml of blood. Pour to sterile plates.

Dextrose Agar

Agar-1.5 gm
Ammonium phosphate ($NH_4H_2PO_4$-0.1 gm)
Potassium chloride (KCl)-0.1 gm
Magnesium sulphate ($MgSO_4$)-0.1 gm
Dextrose-1.0 gm
Distilled water-100ml
pH-7.2-7.4
Agar-2% (Optional)

EMB Agar

Peptone-1.0 gm
Lactose-0.5 gm
Sucrose-0.5 gm
K_2HPO_4-0.2 gm
Eosin-0.5 mg
Methylene blue-6.5 mg
Distilled water-100ml
pH-7.2 to 7.4
Agar-2% (optional)

Endo Agar

Peptone-10.0 gm
Lactose-10.0 mg
K_2HPO_4-3.5 gm
Basic fuschin-0.5 gm
Sodium sulphate-2.5 gm
Distilled water-100ml
pH-7.2 to 7.4
Agar-2%(optional)

LACTOSE BROTH

Beef extract-0.3 gm
Peptone-0.5 gm
NaCl-0.5 gm
Distilled water-100ml
pH-6.7 to 7.0
Agar-2%(optional)

LB medium (Luria Britani)-Sterilization at 121^{0}C for 20 minutes at 15 psi (lb/in^2)

Peptone-1gm
Yeast extract-0.5 gm
NaCl-1.0 gm
Distilled water-100ml
pH-7.2 to 7.4
Agar-2%(optional)

LT Agar

Tryptone: 1.0gm
NaCl: 0.25 gm
Distilled water: 100ml
Agar 2%
Ph -7.0

LT Broth

Maltose: 0.2%

$MgSO_4$:10Mm

Distilled water: 100ml

Malt extract agar

Malt extract-1.5 gm

Agar-2.0 gm

Boil in 100ml of distilled water till the ingredients get dissolved.

Ph-7.2-7.4

Agar-2%(Optional)

MacConkey Agar

Peptone 20.0g

Lactose 10.0g

Bile salts 1.5g

Sodium Chloride 5.0

Neutral Red 0.05g

Crystal Violet 0.001

Agar 15.0g

Distilled water 1.0 liter

pH: 7.4±0.2 (25 °C)

MINIMAL MEDIUM (CONJUGATION)

Ammonium chloride: 10.0 gm

Ammonium nitrate: 2.0 gm

Ammonium sulphate: 4.0 gm

K_2HPO_4: 6.0gm

KH_2PO_4: 3.0gm

$MgSO_{4:}$ 0.2gm

Sucrose: 0.5 gm

Distilled water: 1500ml

Agar-2.5 %, pH 7.2
Threonine, Leucine, Thiamine HCl-20μg/ml stock each
Streptomycine-200 μg/ml

MR-VP Broth

Peptone-7.0 gm
Dextrose-5.0 gm
Potassium phosphate-5.0 gm
Distilled water-100ml
pH-7.2 to 7.4
Agar-2% (optional)

Muller-Hinton Agar

Beef extract-30.0 gm
Casein acid hydrolysate-1.75 gm
Starch-0.15 gm
Distilled water-100ml
pH-7.2-7.4
Agar-2% (Optional)

Nutrient Agar

Beef extract-1 gm
Pepetone-1 gm
NaCl-0.5 gm
Distilled water-100ml
pH-7.2-7.4
Agar-2% (Optional)

Peptone Agar

Agar-1.5 gm
Ammonium phosphate ($NH_4H_2PO_4$-0.1 gm)
Potassium chloride (KCl)-0.1 gm
Magnesium sulphate ($MgSO_4$)-0.1 gm

Dextrose-1.0 gm
Peptone-1.0 gm
Distilled water-100ml
pH-7.2-7.4
Agar-2% (Optional)

Potato Dextrose Agar

Potato-Boiled, smashed-20gm
Dissolve in 100 ml of distilled water, filter
Dextrose-5%, pH-5.6
Agar-2%

Sabouraud's Agar

Peptone-1.0 gm
Glucose-4.0 gm
Distilled water-100ml
pH-7.2 to 7.4
Agar-2% (optional)

S.S agar (Salmonella Shigella Agar)

Beef extract-0.5 gm
Protease peptone-0.5 gm
Lactose-1.0 gm
Bile salts-0.85 gm
Sodium citrate-0.85 gm
Sodium thiosulphate-0.85 gm
Ferric citrate-0.1 gm
Brilliant green-0.033mg
Neutral red-2.5 mg
Distilled water-100ml
pH7.2-7.4
Agar-2% (Optional)

Simon's Citrate Agar

Dihydrogen ammonium phosphate-1.0 gm
K_2HPO_4-1.0 gm
NaCl-5.0 gm
Sodium citrate-2.0 gm
Magnesium sulphate-0.2 gm
Bromothymol blue-0.08 gm
Distilled water-100ml
pH-7.2 to 7.4
Agar-2% (optional)

TBY Agar

Tryptone: 10.0 gm
Yeast extract: 5.0 gm
NaCl: 5.0 gm
Distilled water: 1000ml
pH 7.2 Agar- 2%

Urea Broth

Yeast extract-0.01 gm
Monopotassium phosphate-0.91 gm
Disodium phosphate-0.95 gm
Urea-2.0 gm
Phenol red-0.001gm
Distilled water-100ml
pH-6.8 to7.2
Agar-2% (optional)

Baritte Reagent

Solution A- α-napthol-6.0gm , dissolve in 100ml of 95 % ethanol.

Solution B-16 gm of KOH in 100 ml of distilled water

Mix equal volumes of solution A and solution B at the time of use.

Kovax Reagent-For Indole Test

Amyl alcohol-75 ml

p- Di methyl amine benzaldehyde-5.0 gm

Concentrated HCl-25 ml

Appendix B

BUFFER COMPOSITIONS

1. Acetate Buffer

Solution1:0.2M solution of acetic acid (11.5 ml in 1000ml)

Solution2:0.2M solution of sodium acetate (27.2 g of sodium acetate in 1000ml water)

Mix x ml of solution 1 and y ml of solution 2 and dilute to 100ml

x (ml)	y (ml)	pH
46.3	3.7	3.6
44.0	6.0	3.8
41.0	9.0	4.0
36.8	13.2	4.2
30.5	19.5	4.4
25.5	24.5	4.6

20.0	30.0	4.8
14.8	35.2	5.0
10.5	39.5	5.2
8.8	41.2	5.4
4.8	45.2	5.6

2. Carbonate-Bicarbonate Buffer

Solution1: 0.1 M sodium bicarbonate

Solution2: 0.1 M of anhydrous sodium carbonate

x ml of solution 1 and make up with solution 2. to 100 ml.

x ml	pH
93	9.0
88	9.2
82	94
73	9.6
61	9.8
49	10.0
38	10.2
27	10.4
18	10.6
10	10.8
5.51	1.0

3. Phosphate Buffer

Solution1.0.2M solution of monobasic sodium phosphate

Solution2:0.2 M solution of dibasic sodium phosphate

Take x ml of 1 and y ml of 2 and make up to 200ml with distilled water

x	y	pH
93.5	6.5	5.7
92.0	8.0	5.8
90	10.0	5.9

87.7	12.3	6.0
85	15.0	6.1
81.5	18.5	6.2
77.5	22.5	6.3
73.5	26.5	6.4
68.5	31.5	6.5
62.5	37.5	6.6
56.5	43.5	6.7
51.0	49.0	6.8
45.0	55.0	6.9
39.0	61.0	7.0
33.0	67.0	7.1
28.0	72.0	7.2
23.0	77.0	7.3
19.0	81.0	7.4
16.0	84.0	7.5
13.0	87.0	7.6
10.5	89.5	7.7
8.5	91.5	7.8
7.5	93.0	7.9
5.3	94.7	8.0

TE BUFFER-pH-8.0

1M Tris HCl-2.5 ml, 0.5M EDTA-0.5 ml. Dissolve in distilled water. Adjust pH with HCl. Make up to 250ml with distilled water.

TAE Buffer pH-8.0

50 X-1liter

242 g of Tris base

57.1 ml glacial acetic acid

100 ml-0.5 M EDTA (pH-8.0).Make up to 1000ml with distilled water.

Working solution is 1X.

TBE buffer 5x-Ph 8.0

Tris base-54 gm, Boric acid-27.5gm, 0.5M EDTA-20ml, Adjust pH to 8.0. Make up to 1000ml with distilled water.

Electrophoresis Loading Buffer (6X) - 100ml Contains

0.25%bromophenol blue

0.25%xylene cyanol FF

40% Sucrose in water or 30% glycerol in water (Sore at 4^0C).

Appendix C

ACID DILUTIONS

Acetic Acid

Dilute 116 ml of the concentrated acid to 1 litre of water. (2M)

Hydrochloric Acid

Dilute 172 ml of the concentrated acid to 1 liter with water.(2M)

Nitric Acid

Dilute 122ml of the concentrated acid to 1 liter with water. (2M)

Sulphuric Acid

Pour 56 ml of the concentrated acid slowly and with constant stirring into about 500mls. of water, cool, and dilute to 1 litre (1M)

Sulphurous Acid

Prepare a saturated solution of SO_2 in water: 6-7% by weight.

Dilution of Bases

Ammonia solution concentrated

The commercial product contains about 28% NH_3. (15N)

Ammonia Solution Dilute

Dilute 335 ml of the commercial solution to 1 liter (5N) Barium hydroxide

Shake 70 gm of crystalized salt in 1 liter of water. Filter or siphon off the liquid, and protect from CO_2 of the air 0.2M

Calcium Hydroxide

Shake 2-3 gm of calcium hydroxide with 1 liter of water. Filter or siphon off liquid, and protest from CO_2 of the air. 0.02M

Potassium Hydroxide

Dissolve 112 gm of KOH pellets in water , and dilute to 1 liter. (2M)

Sodium Hydroxide

Dissolve 200 gm pellets in water and dilute to 1 liter. (5M)

Appendix D

CHEMICAL PREPARATIONS

List of commonly used solvents and preparation

Acetic acid (MW 60.05)

Glacial acetic acid is 99.6% (w/v) acetic acid, and is 17.4 M.

1 M : Add 57.5 ml of glacial acetic acid to 800 ml of water and then make to 1 liter with water.

8% (w/v) : Add 80 ml of glacial acetic acid to 800 ml of water and then make to 1 liter with water.

Acetic Acid/ Butanol/Water (15:60:25) For 1000 ml

Combine 150 ml of glacial acetic acid, 600 ml of n-butanol and 250 ml of water.

Aceto-carmine

Add 5.0 grams of orcein to 50 ml of glacial acetic acid. Boil and continue to heat until completely dissolved. Cool and add 50 ml of distilled water. Filter prior to use.

Aceto-orcein

Add 2.0 grams of orcein to 45 ml of glacial acetic acid. Boil and continue to heat until completely dissolved. Cool and add 55 ml of distilled water. Filter prior to use.

Acid Alcohol

Add 1.0 ml of concentrated HCl to 100 ml of 70% (v/v) ethyl alcohol.

Acid Orcinol Reagent (0.1% FeCl in 10% HCl)

Add 0.1 grams of FeCl to 50 ml of 10% (v/v) HCl and make to 100 ml with 10% HCl.

Acryl Amide Solutions

Acryl amide solutions for PAGE are given as total concentration of acryl amide (acryl amide + bisacryl amide) and the amount of cross linker (bisacryl amide). This is listed as the T:C ratio. For example, a 10% gel (10%T:5%C) would contain a total of 10 g rams of acryl amide per 100 ml, and would be composed of 5 grams of acryl amide and 5 grams of bisacryl amide. Usually, a stock solution of 30% acryl amide is produced containing 0.8% bis-acryl amide. Many investigators use 30 grams of acryl amide plus 0.8 gram s of bis-acryl amide per 100 ml of water, but 29.2 grams of acryl amide plus 0.8 grams of bis would be technically correct. In practice, it makes little difference since the gels are diluted to 10% or less. The 30% stock solution is filtered through a 0.45 μ filter and stored at 4 ° C in the dark. For use, the stock solution is diluted with an appropriate buffer (usually a 2X Tris-HCl). The stock solution is stable for about one month. Discard after this period.

Acrylamides in their monomeric form are neurotoxic. Polymerize all acrylamide solutions prior to disposal.

SDS, β-mercaptoethanol and a tracker dye (bromophenol blue) are added at various points. Refer to Chapter 4 for more complete details.

Alcohol Orcinol Reagent (10% orcinol in 95% ethanol)

Dissolve 1.0 gram of orcinol in 95% ethanol to a final volume of 10 ml.

Alkaline Distilled Water

Add one pellet of NaOH to 1 liter of distilled water.

Alkaline Solution for G-banding

Dissolve 2.8 grams of NaOH and 6.2 grams of NaCl to a final volume of 1 liter with water.

ρ-Aminosalicic Acid (PAS MW 175.1)

6% (w/v) Dissolve 6.0 grams of PAS to a final volume of 100 ml with water or buffer.

Ammonium acetate (MW 77.08)

0.1 M Add 7.708 grams of Ammonium acetate to a final volume of 1 liter of water.

Ammonium per sulfate (MW 228.2)

10% (w/v) : Dissolve 1.0 grams of ammonium per sulfate to a final volume of 10 ml with water. Mix fresh, prior to use as a catalyst for PAGE. Normally, about 50 μ l of ammonium per sulfate is added to each 15 of gel solution for polymerization. Dissolve 0.13 grams of Alcian Blue 8GX in 100 ml of water.

Ammonium sulfate (MW 132.14)

2% (w/v)	Add 2 grams of ammonium sulfate to a final volume of 100 ml water.
4.1 M (sat.) 0.001 M	Dissolve 542 grams of ammonium sulfate to a final volume of 1 liter.

n-Amyl alcohol (Pentanol CHOH MW 88.15)

Density = 0.8144 grams/ml

0.38 M : The amyl alcohol can be weighed (33.5 grams) or measured volumetrically by using the density. That is, 33.5 grams 0.8144 grams/ml or 41.1 ml of n-amyl alcohol. Weigh or measure the appropriate amount and dilute to a final volume of 1 liter with water.

Amylase, buffered pH 7.0

1% (w/v) : Dissolve 0.5 grams of amylase to a final volume of 50 ml with 0.01 M sodium phosphate buffer, pH 7.0.

Ascorbic acid (MW 176.12)

2 mM : dissolve 35.2 mg of ascorbic acid to a final volume of 100 ml with water.

ATP (Adenosine triphosphate, MW 507.21)

5 mM : dissolve 254 mg of ATP to a final volume of 100 ml with water or buffer. Dissolve 35.2 mg of ascorbic acid to a final volume of 100 ml with water.

Baker's Formalin

Add 1.0 gram of calcium chloride, 1.0 gram of cadmium chloride and 10 ml of concentrated formalin to 75 ml of water. Make to a final volume of 100 ml with water.

Benzoic acid (MW 122.12)

8 mM : Dissolve 98 mg of benzoic acid to a final volume of 100 ml with water or buffer.

Bis-acryl amide (N,N'-Methylene-bis-acryl amide)

Cross linker for acryl amide gels. Refer to Acryl amide solutions

Biuret Reagent

Add 1.50 grams of CuSO ·HO and 6.0 grams of sodium potassium tartrate to 500 ml of water. Separately make 300 ml of 10% (w/v) NaO H by dissolving 300 grams of NaOH to a final volume of 300 ml with water. Combine the two solutions in a 1 liter volumetric, swirl to mix and make up to 1 liter with water. Store the final solution in a dark, plastic bottle. Discard if black or red precipitate forms.

Bovine Serum Albumin (BSA)

There are many grades of BSA available and care should be taken when using this protein. For routine protein concentration standards, a 96-99% pure fraction may be used. For tissue culture, RIA, or molecular weight standardization, BSA should be obtained which is extracted and purified specifically for that purpose.

1% (w/v) Dissolve 0.5 grams of BSA to a final volume of 50 ml in water or buffer.

Bradford Protein Assay

This procedure uses an absorbance shift in an acidic Coomasie Blue solution. Dissolve 100 mg of Coomasie Brilliant Blue G-250 in 50 ml of 95% ethanol. Add 100 ml of 85% phosphoric acid, and bring to a final volume of 1 liter with distilled water.

Phosphoric Acid is Extremely Corrosive. Handle with Care

Bromophenol blue (Sodium salt, MW 692.0)

0.001 (w/v) : Dissolve 1 mg of Bromophenol blue, sodium salt (Sigma # B7021) to a final volume of 100 ml with either water or buffer.

n-Butanol (C_4H_9OH MW 74.12)

Density = 0.8098 grams/ml

1.1 M : The butanol can be weighed (81.5 grams) or measured volumetrically by using the density. That is, 81.5 grams 0.8098 grams/ml or 100.7 ml of n-butanol. Weigh or measure the appropriate amount and dilute to a final volume of 1 liter with water.

Calcium chloride (MW 110.99)

0.001 M. : Dissolve 0.111 grams of anhydrous calcium chloride to a final volume of 1 liter with water or buffer.

0.08 M. : Dissolve 8.879 grams of anhydrous calcium chloride to a final volume of 1 liter with water or buffer.

2% (w/v). : Dissolve 2 grams of anhydrous calcium chloride to a final volume of 100 ml with water or buffer.

Carnoy fixative

Combine 10.0 ml of glacial acetic acid with 60.0 ml of absolute ethyl alcohol and 30.0 ml of chloroform.

Chloroplast homogenization buffer

To 400 ml of distilled water, add 30.058 grams of sorbitol, 2.23 grams of sodium pyrophosphate, 0.407 grams of magnesium chloride, and 0.176 grams of ascorbic acid. Adjust the pH to 6.5 with HCl and dilute to a final volume of 500 ml.

Chloroplast suspension buffer

To 400 ml of distilled water, add 30.058 grams of sorbitol, 0.372 grams of EDTA, 0.102 grams of magnesium chloride and 5.958 grams of HEPES buffer. Adjust the pH to 7.6 with NaOH and dilute to a final volume of 500 ml.

Citric acid ($H_3C_6H_5O_7 \cdot H_2O$ MW 210.14)

0.1 M : Dissolve 21.01 grams of citric acid to a final volume of 1 liter.

Citrate buffer (Sodium phosphate-Citrate buffer)

0.001 M

pH 4.8 Add 493 ml of 0.2 M Na_2HPO_4 to 507 ml of 0.1 M citric acid.

pH 3.6 Add 322 ml of 0.2 M Na_2HPO_4 to 678 ml of 0.1 M citric acid.

pH 4.2 Add 414 ml of 0.2 M Na_2HPO_4 to 586 ml of 0.1 M citric acid.

pH 5.4 Add 557.5 ml of 0.2 M Na_2HPO_4 to 442.6 ml of 0.1 M citric acid.

pH 6.0 Add 631.5 ml of 0.2 M Na_2HPO_4 to 368.5 ml of 0.1 M citric acid.

pH 6.6 Add 727.5 ml of 0.2 M Na_2HPO_4 to 272.5 ml of 0.1 M citric acid.

pH 7.2 Add 869.5 ml of 0.2 M Na_2HPO_4 to 130.5 ml of 0.1 M citric acid.

pH 7.8 Add 957.5 ml of 0.2 M Na_2HPO_4 to 42.5 ml of 0.1 M citric acid.

Cobaltous Nitrate (MW 182.96)

2% (w/v) Dissolve 2.0 grams of cobaltous nitrate (hexahydrate is very soluble) to a final volume of 100 ml with water. Keep well closed in a cool place.

Coomasie blue (Coomasie Brilliant Blue R250)

0.25% (w/v) 0.001 M : Dissolve 2.50 grams of Coomasie Brilliant Blue R250 to a final volume of 1 liter with 20% (w/v) trichloroacetic acid (TCA). Some investigators use a 0.25% solution of Coomasie Blue in methanol-water-glacial acetic acid (5-5-1).

Copper sulfate ($CuSO_4$ $5H_2O$ MW 249.68)

0.5% (w/v) : Dissolve 0.13 grams of Alcian Blue in 100 ml of water.

0.5% (w/v)

Dissolve 0.5 grams of copper sulfate to a final volume of 100 ml with water.

Copper tartrate carbonate (CTC)

Dissolve 0.5 grams of copper sulfate and 1.0 gram of potassium sodium tartrate to a final volume of 100 ml with water. Combine 1.0 ml of this solution with 50 ml of 2% Na_2CO_3 in 0.1 N NaOH. Must be made fresh, prior to use. Stock solutions are stable.

Crystal Violet

Dissolve 0.1 grams of crystal violet and 0.25 ml of glacial acetic acid to a final volume of 100 ml with water. Prepare a 2% solution filter and store. Dilute and use.

DCMU (3-(3,4-Dichlorophenyl)-1,1-Dimethylurea MW 233.1)

1 x 10_{-4} M 0.001 M. Dissolve 2.3 mg DCMU to a final volume of 100 ml with water or buffer.

5 x 10 M0.001 M Dilute the 1 x 10- M solution 1/200 prior to use.

Dichloro phenolindo phenol (DCPIP MW 290.1)

0.0025 M. : Dissolve 73 mg of DCPIP to a final volume of 100 ml with water or buffer.

0.0001 M. : Dissolve 2.9 mg of DCPIP to a final volume of 100 ml with water or buffer.

Dinitrophenol (DNP MW 184.11)

18.4 mg% Dissolve 18.4 mg of 2,4-dinitrophenol to a final volume of 100 ml with water or buffer.

Diphenylamine reagent

Dissolve 500 mg of diphenylamine in 49 ml of glacial acetic acid. Add 1.0 ml of concentrated HCl.

Dithiothreitol (Cleland's Reagent MW 154.3)

0.01 M. : Dissolve 154 mg of dithiothreitol to a final volume of 100 ml with water or buffer. Dithiothreitol is available from Sigma Chemical Co., St. Louis, Cat # D0632. Dithioerythritol may be substituted.

DOPA (3-(3,4-Dihydroxyphenyl)-L-alanine MW 197.19)

8 mM : Dissolve 158 mg of L-DOPA to a final volume of 100 ml with water or buffer. Note that the maximum solubility of DOPA in water is 165 mg/100 ml (8.3 mM).

EDTA (Ethylenediaminetetraacetic acid MW 292.24)

1 M : Dissolve 292.24 grams of EDTA, free acid to a final volume of 1 liter. If the more soluble disodium salt of EDTA is used, adjust the weight accordingly. The pH can be adjusted with acetic acid or NaOH. For corresponding concentration dilutions, multiply the weight in grams by the desired molarity. For exam ple, for 10 mM EDTA, multiply 292.24 X 0.010 to obtain 2.92 grams of EDTA per liter.

EGTA (Ethylene Glycol-bis(amino ethyl Ether) N,N,N',N'-Tetraacetic Acid MW 380.4)

1 mM : Dissolve 380 mg of EGTA to a final volume of 1 liter with water or buffer.

Eosin

0.5% (w/v) : Dissolve 0.5 grams of Eosin Y in 100 ml of water.

Ethanol (CHOH MW 46.07)

Density = 0.7893 gm/ml

50-95% (v/v). Since 95% ethyl alcohol is less expensive and easier to store than absolute, these dilutions should be made with 95% ethyl alcohol. Unless otherwise stated, denatured alcohol works as well as the more expensive non-denatured. A simple way to make the % solution is to use the appropriate amount of 95% ethanol and dilute to 950 ml instead of 1 liter. For example, to make a 50% (v/v) solution, measure out 500 ml of 95% ethyl alcohol and dilute to a final volume of 950 ml with water. For a 70% solution, measure 700 ml of ethyl alcohol and dilute to 950 ml with water. Absolute ethanol should be used directly as 100% ethanol. It is important for histology that this be truly 100%. Since it is hydroscopic (it absorbs water from the air), do not assume it is absolute unless it is sealed or treated to ensure no water. To test, add a drop to a sample of xylol. If any cloudiness occurs, the alcohol is not absolute.

8.5 M : The ethanol can be weighed (391.6 grams of absolute, 412.2 grams of 95% (v/v) or measured volumetrically by using the density. That is, 391.6 grams 0.7893 grams/ml or 496.1 ml of absolute ethanol. Using 95%, 412.2 grams 0.7893 grams /ml or 522.2 ml. Weigh or measure the appropriate amount and dilute to a final volume of 1 liter with water.

Ethanol-acetic acid fixative for histology (3:1)

To 75 ml of absolute alcohol, add 25 ml of glacial acetic acid. Must be made fresh, just prior to use.

Folin-Ciocalteu Reagent

This is usually purchased premixed, since it is difficult to make. Also known as 2N Folin-phenol reagent.

Giemsa stain

Prepare a stock solution by dissolving 3.8 grams of giemsa powder in 25 ml of glycerin. Heat gently with stirring for about 2 hours at 60 ° C. Cool and add 75 ml of methanol (neutral, acetone free).

For a working solution, dilute the stock solution 1/10 with water before use.

For chromosome banding, combine 5.0 ml of stock Giemsa, 3.0 ml of absolute methanol, 3.0 ml of 0.1 M citric acid and 89 ml of distilled water. Adjust the pH of the solution to 6.6 with $NaHPO_4$.

Glucose (MW 180.16)

10% (v/v) : Dissolve 10 grams of D-glucose (dextrose) in a final volume of 100 ml with water or buffer.

Glutaraldehyde (GTA)

5 % : GTA is usually supplied as a 25% or 50% (w/v) solution. It is used for electron microscope fixation as a 5% solution in a buffer. For routine use, add 20 ml of 25% GTA to 80 ml of 0.2 M sodium cacodylate buffer, pH 7.4.

Glycerol (MW 92.09)

10% (v/v) : To 10 ml of glycerol (glycerin) add enough water to make a final volume of 100 ml.

8 M : Weigh 73.67 grams of glycerol and add to a final volume of 100 ml. alternatively, measure 499.1 ml of glycerol and make to a final volume of 1 liter (the density of glycerol at room temperature is 1.476) with water or buffer.

For 8 M glycerol in MT buffer, make a 2X MT buffer for use as the diluent.

Glycine (MW 75.07)

0.192 M. : Dissolve 1.44 grams of Glycine to a final volume of 100 ml with water or buffer.

Gram's iodine

Dissolve 0.33 grams of iodine and 0.67 grams of potassium iodide to a final volume of 100 ml with water.

HEPES (N-[2-Hydroxyethyl] piperazine-N'-[2-ethanesulfonic acid] MW 238.3)

50 mM : Dissolve 11.92 grams of HEPES, free acid to a final volume of 1 liter. If hemi sodium salt is used, adjust weight accordingly. Do not use sodium salts unless specified. Hemi sodium salt contains 0.5 moles of sodium for each mole of HEPES.

10 mM

pH7.6. Dissolve 2.38 grams of HEPES, free acid in 900 ml of water. Adjust the pH with NaOH or HCl to 7.6. Adjust the final volume to 1 liter with distilled water.

Hydrochloric Acid (HCl MW 36.46)

Concentrated HCl has a molarity of approximately 11.6.

HCl is a gas, which is soluble in water and which comes in the form of concentrated reagent grade HCl. This solution is approximately 36-38% (w/v) HCl. To make a 1 N solution, add 86 ml of concentrated HCl to 800 ml of water and dilute to a final volume of 1 liter. For 0.1 N, dilute the 1 N by a factor of 10.

For % solutions, note that liquid HCl is only 38% HCl, thus a 1% solution would require 2.6 ml of concentrated HCl (1/.38) per final volume of 100 ml.

Janus Green B

0.01% (w/v) : Dissolve 10 mg of Janus Green B in 2-3 ml of absolute ethanol. Dilute to a final volume of 100 ml with water.

Knudson Media

Knudson X4 Stock

$Ca (NO_3) \cdot 4H_2O$	4.0 grams
$(NH_2) SO_4$	2.0 grams

$MgSO_4 \cdot 7H_2O$	1.0 gram
Distilled H_2O	1.0 liter
B5 Minor Elements:	
H_2SO_4	0.5 ml
$MnCl_2 \cdot 4H_2O$	2.5 grams
H_3B_3O	2.0 grams
$ZnSO_4 \cdot 7H_2O$	50 mg
$CoCl_2 \cdot 6H_2O$	30 mg
$CuCl_2 \cdot 2H_2O$	15 mg
$Na_2MoO_4 \cdot 2H_2O$	25 mg
Distilled H_2O	1.0 liter
Ferric Citrate	
$FeC_6H_5O_7 \cdot 5H_2O$	2.5 grams
Distilled H_2O	100 ml
Stock Phosphate:	
K_2HPO_4	25 grams
Distilled H_2O	100 ml

1X Media : Add 250 ml of Knudson X4 to 750 ml of distilled water. Add 0.5 ml of B5 Minor Elements, 0.5 ml of Stock Phosphate, and 0.4 ml of Ferric citrate. Adjust the pH to 5.5 with HCl, add 15 grams of agar and heat to dissolve. Autoclave and pour into plates.

Note: 2.50 grams of sucrose may be added prior to adjustment of the pH, if desired. It is not necessary for germination of spores, but adds an organic source for mutants and abnormal fern growths.

Lacto phenol cotton blue

20% phenol (as disinfectant), 0.05%cotton blue(acid dye and stains the chitin), 20% lactic acid in glycerol (40ml) (clears tissue and preserves fungal structure).Make up to 100ml with distilled water.

LPS buffer (Lower Pad Solution buffer)

Dissolve 1.5 grams of KCl, 0.5 grams of MgCl, and 0.5 grams of streptomycin sulfate in 500 ml of water. Add 40 ml of 1 M phosphate buffer, pH 6.5 and dilute to 1 liter with water.

Magnesium chloride ($MgCl_2$ MW 95.23)

1 mM. Dissolve 95.2 mg of magnesium chloride per final volume of 1 liter.

4 mM. Dissolve 0.381 grams of magnesium chloride per final volume of 1 liter.

10 mM. Dissolve 0.952 grams of magnesium chloride per final volume of 1 liter.

0.1 M. Dissolve 9.523 grams of magnesium chloride per final volume of 1 liter.

Note: A single stock solution of 1 M $MgCl_2$ can be mixed by dissolving 95.23 grams of magnesium chloride to a final concentration of 1 liter with water, and all dilutions made appropriately from this stock solution.

Magnesium sulfate (Mg SO MW 120.39)

5% (w/v) Dissolve 5.0 grams of magnesium sulfate to a final volume of 100 ml with water or buffer.

Mayer's hematoxylin

Purchase commercially

A. Dissolve 1.0 gram of hematoxylin in 10 ml of absolute ethanol. Dissolve 20 grams of potassium alum (KAl (SO) · $12H_2O$) in 200 ml of water. In a chemical hood, with protection against explosion, bring the potassium alum solution to a boil and add hematoxylin/ethanol mixture. Continue to boil for approximately 1 minute. Add 0. 5 grams of mercuric oxide and cool rapidly. Add 0.5 ml of glacial acetic acid. Filter before use. This mixture is stable for about two months.

B. *Alternatively* : Dissolve 5.0 grams of hematoxylin in 50 ml of absolute ethanol and add to 650 ml of warm water. Heat gently until the hematoxylin dissolves and then add 300 ml of glycerin, 0.3 grams of sodium iodate and 20 ml. of glacial acetic acid. Co ol and make volume up to 1 liter with distilled water. Filter before use.

β-Mercaptoethanol (MW 78.13) 0.5 M : Density = 1.2 grams/ml. Use either 3.91 grams OR 3.26 ml of mercaptoethanol in a final volume of 100 ml of water or buffer.5% (w/v) : Use 5.0 grams or 4.167 ml in a final volume of 100 ml of water or buffer.

MES (2-(N-Morphilino) ethanesulfonic acid MW 195.2)

0.1 M : Dissolve 1.952 grams of MES to a final volume of 100 ml with water or buffer.

Methanol (CH_3OH MW 32.04)

Density = 0.7914 grams/ml

22 M : The methanol can be weighed (704.9 grams) or measured volumetrically by using the density. That is, 704.9 grams 0.7914 grams/ml or 890.7 ml of methyl alcohol. Weigh or measure the appropriate amount and dilute to a final volume of 1 liter with water.

Methanol/Acetic Acid (for fixing proteins in acryl amide gels)

45%:12% Add 120 ml of glacial acetic acid to 450 ml of methanol and dilute to a final volume of 1 liter with water.

Methanol/Acetic Acid (for destaining or fixing proteins in acryl amide gels)

5%:7% : Add 70 ml of glacial acetic acid to 50 ml of methanol and dilute to a final volume of 1 liter with water.

Methyl green

0.2% (w/v) : Dissolve 0.2 grams of methyl green to a final volume of 100 ml with 0.1 M acetate buffer, pH 4.2.

Acetate buffer (0.1 M pH 4.2) is prepared by dissolving .361 grams of sodium acetate (trihydrate) in approximately 80 ml of water. Add .42 ml of glacial acetic acid and adjust the volume to 100 ml with water.

Microtubule buffer (MT buffer)

Dissolve 19.52 grams of MES in 800 ml of distilled water. Add 0.380 grams of EGTA and 47.62 grams of MgCl. Adjust the pH to 6.4 with HCl or NaOH and dilute to a final volume of 1 liter with distilled water.

Minimum essential medium (MEM)

For all purposes of this manual, MEM refers to Eagle's MEM. While it is possible to mix this medium, it is infinitely easier (and less expensive) to purchase the media pre-mixed from any number of commercial sources.

ρ-Nitrophenyl phosphate (MW 263.1)

0.05 M. Dissolve 1.32 grams of -nitrophenyl phosphate to a final volume of 100 ml with water or buffer.

0.8% (w/v) : Dissolve 0.8 grams of -nitrophenyl phosphate to a final volume of 100 ml of water or buffer.

Perchloric acid (PCA MW 100.47)

2% (w/v)

Dissolve 2.0 grams of PCA to a final volume of 100 ml with water or buffer.

Percoll

Colloidal PVP coated silica.

Periodic Acid (Periodate Used for PAS reaction)

Dissolve 0.6 grams of periodic acid in 100 ml of water and add 0.3 ml of concentrated nitric acid.

Phenol mixture

Combine 555 ml of aqueous phenol (or 500 grams of phenol crystals plus 55 ml of water) with 70 ml of mcresol. Add 0.5 grams of 8-hydroxyquinoline.

Phenol will cause severe burns and readily dissolves all plastic and rubber compounds. Use extreme caution when handling this compound.

Phosphate buffered saline (PBS)

Mix 100 ml Ca/Mg fr ee 10X PBSA with 800 ml of distilled water. Separately, dissolve 0.1 gram of magnesium chloride and 0.1 gram of anhydrous calcium chloride to a final volume of 100 ml with water. With constant stirring, slowly add the magnesium/calcium chloride solution t o the diluted PBSA. If a precipitate forms, start over, and add slower with continuous stirring.

Ca/Mg free Phos phate buffered saline - 10X (10X PBSA)

Dissolve 80 grams of NaCl, 2.0 grams of KCl, 15.0 grams of Di basic sodium phosphate and 2.0 grams of Monobasic potassium phosphate in 1 liter of distilled water. This makes a 10X solution of Ca/Mg free phosphate buffered saline. Dilute 1:10 prior to use. Store in a refrigerator.

Phosphate buffered saline-Tween 20 (4.8)

Mix PBS and add 0.1% (v/v) Tween 20.

Phytohemaglutinin (PHA)

Available as kidney bean lectin. It is typically used as a stock solution of 10-20 g/ml in balanced salt solution. For tissue culture it must be cold sterilized prior to use.

Potassium chloride (KCl MW 74.55)

1 M : Dissolve 74.55 grams of KCl to a final volume of 1 liter with water or buffer. For other concentrations, multiply the weight by the required molarity. For example, for 0.150 M (150 mM), use 0.150 X 74.55, or 11.183 grams of KCl in 1 liter of water or buffer.

Potassium cyanide (KCN MW 65.11)

8 mM : Dissolve 52 mg KCN to a final volume of 100 ml with water or buffer.

Potassium phosphate, monobasic (KHPO MW 136.09)

0.01M. : Dissolve 1.36 grams of monobasic potassium phosphate to a final volume of 1 liter with water.

Potassium phosphate, dibasic (KHPO MW 174)

0.01 M. : Dissolve 1.74 grams of dibasic potassium phosphate to a final volume of 1 liter with water.

Potassium phosphate buffer

0.01M

pH 7.4 : Prepare 500 ml of 0.01 M KH_2PO_4 and 500 ml of 0.01 M KH_2PO_4 Place the K_2HPO_4 onto a magnetic stirrer and insert a pH electrode. Add the KH_2PO_4 slowly to adjust the pH to 7.4.

Potassium hydroxide (KOH MW 56.10)

0.5 N : Dissolve 28.05 grams of KOH to a final volume of 1 liter with water.

10% (w/v) : Dissolve 10 grams of KOH to a final volume of 100 ml with water. Store in a plastic container.

n- Propanol (CHOH MW 60.11)

Density = 0.8035 grams/ml

3 M. The n-propanol can be weighed (180.3 grams) or measured volumetrically by using the density. That is, 180.3 grams 0.8035 grams/ml or 224.4 ml of n-propanol. Weigh or measure the appropriate amount and dilute to a final volume of 1 liter with water.

Protein Buffer

Dissolve 1.46 grams of KHPO and 0.92 grams of K2HPO in 80 ml of distilled water. Add 2.5 grams of crystalline serum albumin and adjust the volume to a final 100 ml with water.

Ribonuclease

0.1% (w/v) : Dissolve 10 mg of pancreatic ribonuclease type A in 10 ml of water or buffer. Use for enzyme treatment of histological sections by floating 0.5-1.0 ml of this solution onto the section, with the slide set into a covered petri plate.

Safranin

Dissolve 2.5 grams of Safranin O in 10 ml of 95% ethanol and dilute to 100 ml with water.

Saline (NaCl)

0.85% (w/v) : Saline refers to a solution of NaCl, with the most common usage for that which is isotonic to mammalian blood cells, notable a 0.85% or 0.9% solution. To mix, dissolve 8.5 grams of NaCl to a final volume of 1 liter with water.

Saline citrate (1/10 dilution of SSC)

Dissolve 0.878 grams of NaCl and 0.294 grams of sodium citrate to a final volume of 1 liter with water.

Saline citrate buffer (SSC)

20X : It is common to prepare this buffer as a stock 20X solution, to be diluted to 2X, 1X or 0.1X prior to use. To prepare a 20X stock solution, dissolve 175 grams of NaCl and 88 grams of sodium citrate in 900 ml of water. Adjust the pH to 7.0 with 1 N HCl and bring to a final volume of 1 liter.

For use, as a 1X SSC, dilute 1 part 20X stock with 19 parts distilled water. For a 2X SSC, dilute 1 part 20X stock with 9 parts water.

Schiff's Reagent

Dissolve 0.8 grams of basic fuschin in 85 ml of distilled water. Add 1.9 grams of sodium meta bisulfite and 15.0 ml of 1 N HCl. Place the solution in separating funnel and shake at 2 hour intervals for a period of approximately 12 hours. Add 200 grams of activated charcoal, shake for 1 minute and filter the clear solution. If the solution is still pink, add another 100 grams of charcoal and shake for an additional minute. Filter and store in a dark bottle. Solution should be clear (no pink coloration) for use.

Scott solution

Dissolve 2.0 grams of sodium bicarbonate and 20.0 grams of magnesium sulfate in water to a final volume of 1 liter. Add a pinch of thymol to retard the growth of molds.

SDS

Refer to Sodium lauryl sulfate.

1X SDS-Electrophoresis Running Buffer

2X SDS Sample Buffer

Dissolve 1.52 grams of Tris base, 2.0 grams of SDS, 20 ml of glycerin, 2.0 ml of -mercaptoethanol and 1 mg of bromophenol blue to a final volume of 100 ml with water.

Silver nitrate solution (for electrophoresis staining)

Dissolve 0.15 grams of NaOH in 150 ml of water. Add 3.5 ml of concentrated NHOH and bring to a volume of

200 ml. Separately, dissolve 2.0 grams of silver nitrate in a final volume of 10 ml. With constant stirring, add 8.0 ml of the silver nitrate to the 200 ml of NaOH/NHOH.

This solution should be prepared immediately prior to use, and used within 30 minutes.

Dispose of this solution with copious flushing. It becomes explosive upon drying.

SM agar medium (Slime Mold medium of Sussman)

Dissolve 10.0 grams of glucose, 10.0 grams of peptone, 1.0 gram of yeast extract, 1.0 gram of MgSO, 1.5 grams of KHPO, 1.0 gram of K2HPO and 20.0 grams of agar to a final volume of 1 liter. Heat to dissolve the agar, autoclave and dispense to petri plates.

Sodium acetate (MW 82.04)

1 M. : Dissolve 82.04 grams of sodium acetate to a final volume of 1 liter with water or buffer.

0.02 M. : Dissolve 1.64 grams of sodium acetate to a final volume of 1 liter with water or buffer.

Sodium acetate buffer

1 M pH 5.7 : To 925 ml of 1 M sodium acetate, add 75 ml of 1 M acetic acid.

Sodium barbitol

0.2% (w/v) : Dissolve 0.2 grams of sodium barbitol to a final volume of 100 ml with water.

Sodium bicarbonate (NaHCO MW 84.0)

0.1 M : Dissolve 0.84 grams of NaHCO to a final volume of 100 ml with water or buffer.

Sodium carbonate (MW 106.0)

2% (w/v) Dissolve 2.0 grams of sodium carbonate to a final volume of 100 ml with 0.1 N NaOH.

0.1 N NaOH : Used for Lowry Protein assay.

Sodium chloride (MW 58.44)

M For a molar solution of sodium chloride, dissolve 58.44 grams of NaCl to a final volume of 1 liter with water or buffer. For corresponding dilutions, multiply the weight by the molarity required. For example, for 0.05 M, multiply 58.44 by 0.05 or 2.92 grams/liter.

% For % solutions, they are invariably w/v. For a 1 % (w/v) solution, dissolve 1.0 gram of NaCl to a final volume of 100 ml with water or buffer. Multiply the weight by a corresponding change in % for other concentrations.

200,300,400 mOsM : Osmoles for NaCl are calculated as twice the molar concentration. Thus, a 200 mOsM solution would be .100 M NaCl. Likewise, 300 mOsM would be .150 M and 400 mOsM would be .200 M NaCl.

Sodium citrate (MW 294.10)

0.09 M. Dissolve 2.65 grams of sodium citrate to a final concentration of 100 ml with water.

Sodium citrate/formaldehyde (for silver stained proteins)

Dissolve 5.0 grams of sodium citrate in 800 ml of water, add 5.0 ml of concentrated formalin (37% formaldehyde solution) and dilute to 1 liter with water.

Sodium deoxycholate (Deoxycholic acid, sodium salt MW 392.58)

0.15% (w/v) : Dissolve 150 mg of deoxycholic acid, sodium salt to a final volume of 100 ml with water.

Sodium fluoride (NaF MW 42.0)

0.1 M : Dissolve 4.2 grams of NaF to a final volume of 1 liter with water.

Sodium lauryl sulfate (SDS or SLS MW 288.38)

0.1% (w/v) : Dissolve 0.1 grams of SDS to a final volume of 100 ml with water or buffer. Mix by gentle stirring, do not shake.

10% (w/v) : Dissolve 10 grams of SDS to a final volume of 100 ml with water.

SDS should not be inhaled in its powder form. When weighing, use a mask, or better, a hood.

Sodium phosphate, monobasic (NaHPO HO MW 137.99)

1 M. : Dissolve 14.01 grams of sodium perchlorate to a final volume of 100 ml with water or buffer.

0.01 M. : Dissolve 1.38 grams of monobasic sodium phosphate to a final volume of 1 liter.

Sodium phosphate, dibasic (NaHPO · 7HO MW 268.07)

1 M. : Dissolve 268.07 grams of dibasic sodium phosphate to a final volume of 1 liter.

0.2 M : Dissolve 53.61 grams of dibasic sodium phosphate to a final volume of 1 liter.

0.01 M : Dissolve 2.68 grams of dibasic sodium phosphate to a final volume of 1 liter.

Sodium phosphate buffer

These are the most common buffers used in biology. They are produced by adding equimolar solutions of $KHPO_4$ and $NaHPO_4$. Equal volumes of the two will yield a pH of 7.0, while sodium phosphate will increase the pH. Increased volumes of potassium phosphate will decrease th e pH. The pH can be adjusted from 5.4 to 8.2.

If a pH of 7.0-8.2 is desired, start with 500 ml of sodium phosphate and add potassium phosphate while stirring and monitoring the pH with a pH meter until the desired pH is reached.

If a pH of 5.4-7.0 is desired, start with 500 ml of potassium phosphate and add sodium phosphate until the desired pH is reached.

Typically, the molarity of the buffer will range from 0.01 to 0.1 M. Use the appropriate molarity of $KHPO_4$ and $NaHPO_4$. That is, if 0.05 M buffer is desired, use 0.5 M $KHPO_4$ and 0.5 M $NaHPO_4$ as directed above.

Sodium potassium phosphate buffer

Refer to Sodium phosphate buffer.

Sodium pyrophosphate ($Na_2PO_4 \cdot$ 10HO MW 446.06)

10 mM : Dissolve 0.446 grams of $Na_2PO_4 \cdot 10HO$ to a final volume of 100 ml with water.

Sodium succinate (MW 270.16)

0.6 M : Dissolve 16.2 grams of succinic acid, sodium salt to a final volume of 100 ml with water or buffer.

Sorbitol (MW 182.17)

0.33 M : Dissolve 60.12 grams of sorbitol to a final volume of 1 liter with water or buffer.

0.2 M pH 7.5 : Dissolve 24.14 grams of $NaHPO_4$ and 4.08 grams of $KHPO_4$ in 800 ml of water. Dilute to a final volume of 1 liter.

Sucrose (MW 342.3)

1.0 M : Dissolve 34.2 grams of sucrose to a final volume of 100 ml with water or buffer. For other molarities, multiply the weight by the required molar concentration. For example, for 0.25 M sucrose, weight 34.2 x 0.25 or 8.55 grams to a final volume of 100 ml.

40% (w/v) : Dissolve 40 grams of sucrose to a final volume of 100 ml with water or buffer. Dilute this solution for lower percent requirements. If using for sucrose density gradients, the sucrose should have 0.1 ml of diethylpyrocarbonate added, the solution brought to a boil for 3-5 minutes and cooled before use. This will eliminate RNAase, which would otherwise be a contaminant of the solution. Store all sucrose solutions in a refrigerator.

Sulfuric Acid (H_2SO_4 MW 98.08)

Caution: Sulfuric acid is extremely caustic and will cause severe burns. It must always be added to the water, when making dilutions. Upon addition to water or alcohol, heat will be generated while the solution will contract in volume. Use extreme care in handling this acid.

Concentrated H_2SO_4 is 17.8 M or 35.6 N. 1.09 N Add 30.6 ml of concentrated sulfuric acid slowly, with constant stirring, and with adequate protection from splashes, to approximately 800 ml of water. Cool and make up volume to 1 liter with water.

Sulfurous acid (for Feulgen Reaction)

Add 1.0 ml of concentrated HCl and 0.4 grams of sodium bisulfite to 100 ml of distilled water. This solution should be made fresh prior to use. It does not store well.

Swabbing detergent

For tissue culture purposes, use a non-toxic detergent designed for surgical scrubbing. e.g. Phisohex, Betadine or equivalent. For most routine swabbing, 70% (v/v) ethanol is sufficient and has the advantage that it will leave no residue.

TEMED (N,N,N',N'-tetramethylethylenediamine)

Catalyst for PAGE. Use directly and add 10 l TEMED per 15 ml of gel solution.

Trichloroacetic acid (TCA CCl_3COOH MW 163.4)

Extremely caustic acid. Handle with care.

72% (w/v): Dissolve 72 grams of TCA to a final volume of 100 ml. TCA is hydroscopic and will readily absorb water. The solid crystals will become liquid if the stock bottle is placed in warm water, with a loose cap (melting point 57-58 ° C. It is easier to handle as a liquid. Storage of solutions greater than 30% (w/v) are not recommended as decomposition is rapid. Therefore these solutions should be made as needed.

Tris buffer

There are many variations on the basic Tris-HCl buffer combination, most of which are commercially available. Solutions with EDTA are known as TE buffers, while solutions with EDTA and acetic acid are known as TAE buffers. The terminology varies with the author, with Tris buffer being used to mean Tris-HCl solutions.

The basic buffer is a combination of Tris (tris (hydroxymethyl) aminomethane) and HCl acid. These are sometimes referred to as Tris-base and Tris-HCl solutions. Tris buffers should not be used below a pH of 7.2 or above a pH of 9.0. Tris buffers are also ex tremely temperature senstive. Directions are given for room temperature (25 ° C). The pH will decrease approximately 0.028 units for each degree decrease in temperature.

1 M : Dissolve 121 grams of Tris in 800 ml of distilled water. Adjust the pH with concentrated HCl. Dilute to a final volume of 1 liter. Lower required molarities can be diluted from this stock or mixed as combinations of lower molarities of Tris and HCl. It is important to measure the pH at the temperature an d molarity that will be used in the final analysis.

Tris -Glycine buffer

5XDissolve 15.1 grams of tris base and 72.0 grams of glycine to a final volume of 1 liter. For use, dilute 1 part 5X buffer with 4 parts water.

Trypan blue

0.2 % (w/v) : Dissolve 0.2 grams of trypan blue to a final volume of 100 ml with water.

Trypsin

0.25% : 0.25% Dissolve 0.25 grams of crude trypsin in PBSA to a final volume of 100 ml. Cold sterilize by filtration.

Alternatively, purchase pre-diluted crude trypsin, sold as 1:250 which is pre-sterilized as well.

Note : When using trypsin for tissue disaggregation, it must be subsequently inhibited by the use of serum in the culture media, or by the addition of soya bean trypsin inhibitor.

Trypticase soy broth

Add 17.0 grams of trypticase peptone, 3.0 grams of phytone peptone, 5.0 grams of sodium chloride, 2.5 grams of dipotassium phosphate and 2.5 grams of glucose to 1 liter of water. Adjust the pH to 7.3, and autoclave.

Tween 20 or 80 (Polyoxyethylene sorbitan mono-oleate)

1% (v/v) Add 1.0 ml of Tween to 90 ml of water. Mix and dilute to a final volume of 100 ml with water. Note that Tween is extremely viscous and care must be taken to accurately pipette 1.0 ml. Wipe the outside of the pipette before dispensing.

Urea (MW 60.06)

2.5 M : Dissolve 15.02 grams of urea to a final volume of 100 ml with water or buffer.

10 M : Dissolve 60.06 grams of urea to a final volume of 100 ml with water or buffer.

14 M : Dissolve 84.08 grams of urea to a final volume of 100 ml with water or buffer.

Appendix E

SPRAY REAGENTS

Aldehydes-Fuchsin
Alkaloids- Dragendorff's
Amines-Ninhydrin
Amino acids and related compounds-Ninhydrin
Antioxidants-Phosphomolybdic acid
Barbiturates-Mercuric diphenylcarbazone
Flavanoids-Antimony trichloride

Glycolipids -Bial's Orcinol, Diphenylamine
Lipids-Antimony trichloride, Bial's Orcinol, Bromothymol blue, Cupric acetate
2, 7-Dichlorofluorescein
Diphenylamine
Dragendorff's
Molybdenum blue
Ninhydrin

Potassium- Dichromate/H2SO4.

Rhodamine B

Rhodamine 6G

Phenols-Phosphormolybdic acid

Rhodamine B

Steroids-Antimony pentachloride

Antimony trichloride

Phosphormolybdic acid

Terpenes-Antimony pentachloride

VIitamins-Antimony pentachloride

Iodoplatinate

Charring reagent-Potassium dichromate/H_2SO_4

Preparation of Reagents

Ammonia

25% w/v in water. For neutralizing ninhydrin fixer for chromatograms and numerous other applications.

Aniline-Diphenylamine

1.8% w/v aniline and 1.8% w/v diphenylamine in acidified acetone.

For detecting reducing sugars.

Antifoam

25% antifoam reagent in methylene chloride.

For preventing/dispersing foam.

Antimony Trichloride

22% w/v antimony trichloride in chloroform.

For detecting vitamin A, steroid glycosides and related compounds.

Bromocresol Green

0.04% w/v bromocresol green in ethanol.

For detecting organic acids and bases.

Chloramine-T

5% of reagent in 0.5% sodium hydroxide.

For detection of phenolic compounds.

Dragendorffs reagent.

0.11% potassium iodide and 0.6millimole bismuth subnitrate in 3.5M acetic acid.

For detecting alkaloids and quaternary nitrogen compounds.

Ferric chloride-potassium ferricyanide

3% solutions in 2M hydrochloric acid

For detecting aromatic amines and phenolic compounds.

Ferric chloride

2.7% w/v of salt in 2M hydrochloric acid.

For detecting phenols and phenolic acids.

Fluorescamine

0.005% of the reagent in acetone.

For detecting primary amino acids and amines.

Molybdenum blue

1.3% w/v molybdenun oxide in 4.2M sulphuric acid.

For detecting phospholipids and related compounds.

Ninhydrin

0.2% w/v ninhydrin in ethanol.or in 6% acetone.

For detecting amino acids, amines, and amino sugars. (amphetamines)

Ninhydrin Fixer reagent

Acidified ethanol containing 1% saturated cupric nitrate.

Fixative for ninhydrin chromatograms.

Orcinol Ferric Chloride (Bial's reagent)

0.9% w/v ferric chloride and 0.55% w/v orcinol in acidified ethanol.

For detecting sugars,glycosides, and sulfolipids.

Oxime (8-hydroxyquinoline)

0.25g in 8:20 ethanol/water

For detection of cations, AL, Cr, Fe, Co, Ni, Mn, Zn.

Phosphomolybdic acid

5% of reagent in Propan 2-ol

For detection of Fatty acid methyl esters, keto acids, lipids, steroids, and sulphated bile acids.

Phosphoric acid / cupric acetate

3% cupric acetate w/v in 8% phosphoric acid v/v.

For detection of Phospholipids.

Rhodamine B

0.25% of reagent in ethanol.

For detecting lipids, and

For detecting metals (Au, Bi, Cd, Fe, Hg, Mo, TI, V, and W.)

Rhodamine 6G

0.01% w/v rhodamine 6G in methylene chloride.

For detecting lipids.

Sulphuric acid spray reagent

5% w/v of the acid in ethanol.

Charring reagents for organic compounds. For detection of bile acids

Appendix F

ACIDS AND BASES
Concentrated Acids

Acids	Molarity	Percent	S.G.	1M soln
Acetic acid MW=60.05	17.4M	99%	1.05	58 ml / liter
HCl MW=36.46	11.6M	36%	1.18	Use 86 ml / liter
Nitric acid MW=63.01	16.4M	69%	1.42	61 ml / liter
H_2SO_4 17.6M MW=98.07	36N	95%	1.84	56 ml / liter (1M) 28 ml / liter (1N)
Ammonia solution MW=35.05	14.5M 15N	28%	0.88	71 ml/ liter (1M)
Hydrobromic acid	9 N	48%	1.49	11.1 ml /liter (1M)
Hydrofluoric acid MW=20.01	27M	48%	1.29	37 ml /liter (1M)

Hydriodic acid	7M	57%	1.70	143 ml / liter (1M)
Per chloric acid				
MW=100.46	11.7M			
9.5M	70%			
61%	1.71			
1.66	85.5mls / liter (1M)			
106 ml / liter (1M)				
Phosphoric acid				
MW=97.99	45N			
	15M	88%	1.69	22.5 ml / liter (1N) 68 ml / liter (1M)

References

A text book of Biochemistry: (1966),Edward Staunton West ,Wilbert R. Todd, Howard S. Mason, John T. Vanberuggen, IV Edition, Oxford and IBH publishing Co. Pvt. Ltd., New Delhi

Analysis of total phenols and other compounds: (1999), Singleton V L., e.t. al., Journal of Biological sciences, 6(2),249-252

Animal cell culture methods: (2006),Volume 57, Jennie P. Mather, David Barens, Academic press, USA.

Antioxidant activity and total phenolic content of Iranian Ocimum accessories: (2003), Javanamardi J., Vivanco, J M., Food chemistry, 83, 547-550

Antioxidant activity of the leaves of Cydonia vulgaris: (2001), Ali Y et al., Turk J Med Sci., 31, 23-27

Antioxidant determination by the use of a stable free radical: (1958), Blois MS, Nature, 181, 1, 1199-1200

Biochemistry: Zubay

Biotechnology of Secondary Metabolites: (1995), Ramawat et .al.

Chromatography: A laboratory Hand Book of Chromatographic and Electrophoretic Methods (1975), Erich Heftman, III Edition, Van Nostrand Reinhold Company, New York.

Culture of Animal cells: A manual of basic techniques: (1994), Ian Freshney, R., Willey-Liss Publication, New York.

Industrial Microbiology: (2003), Cassida, L. E., JR, New Age International Publishers, New Delhi.

Introduction to practical Molecular Biology: (1998), Dabre, P D, John Willey and Sons LTD., New York

Laboratory Techniques in Biochemistry and Molecular Biology: Work and Work.

Methods in Enzymology, A Guide to Molecular Cloning Techniques: Vol 152 (1996), Berger et al., Academic Press, Sandiago.

Microbiology A Human Perspective: (1998), Nester E W, Nancy N. Pearsall, Densie G.Anderson, Martha T. Nester., WCB McGraw-Hill, Boston.

Microbiology Applications (A laboratory Manual in General Microbiology): Benson H J., Brown Publications

Microbiology Laboratory Manual: Cappuccino et al.

Microbiology: (2007) Pelzcar, J, Chan, E C S, Noel R Kreig, Tata McGraw-Hill Publishing Company Limited, New Delhi.

Microbiology: Prescott

Molecular cloning: A Laboratory Manual: (2000), Maniatis, T, Fritsch, E F and Sambrook, J, Cold Spring Harbour Laboratory, New York.

Molecular Cloning: A Laboratory Manual Sambrook, J and Russell, D. (2000) 3rd ed., Cold Spring Harbor Laboratory.

Phytochemical methods: (1976), Harborne, J B, Chapman and Hall, London.

Plant Biochemistry and Biotechnology: (1999), Peter J. Lea, Richard C. Leegood, John Wiley and Sons, New York.

Plant Biotechnology: Slater, Scott, and Flower.

Plant Cell and Tissue Culture: Narayanaswamy.

Plant Cell Culture, A Practical Approach: Dixon et al., IRL Press., Oxford.

Plant Cell Tissue and Organ Culture: Bajaj, Y P S et al.,

Plant Tissue Culture, Bhojwani et al.: Elsevier, Amsterdam.

Plant Tissue Culture: (1985), Second edition, John H. Dodds, and Lorwin W. Roberts, Cambridge University Press, Cambridge.

Principles of Biochemistry: (1990), Albert A. Leghninger, CBS Publishers and Distributors Pvt. Ltd. New Delhi.

Principles of Gene Manipulation: Primrose, S. B., Twyman, R. M. and Old, R.W. 2001. Blackwell, Oxford.

Sex Determination of In Vitro Developed Buffalo Embryos by DNA Amplification: (1993), K.B.C. Appa Rao et al., Molecular reproduction and Development, 36, 291-296

Thin Layer Chromatography: A Laboratory Hand Book, (1969) Edited by Egon Stahl, George Allen and Unwin Ltd., London.

Tools of Biochemistry: Cooper, T G.

http://www.ncbi.nlm.nih.gov

http://www.frodo.wi.mit.edu

Index

F

G

H

I

L

M

N

O

P

Q

R

S

T

V

W

Y

MIX
Papier | Fördert gute Waldnutzung
FSC® C083411

Zeitfracht Medien GmbH
Ferdinand-Jühlke-Straße 7
99095 Erfurt, Deutschland
produktsicherheit@kolibri360.de